Neuroscience for Coaches

How to use the latest insights for the benefit of your clients

Amy Brann

KoganPage

LONDON PHILADELPHIA NEW DELHI

First published in Great Britain and the United States in 2015 by Kogan Page Limited

2nd Floor, 45 Gee Street	1518 Walnut Street, Suite 1100	4737/23 Ansari Road
London EC1V 3RS	Philadelphia PA 19102	Daryaganj
United Kingdom	USA	New Delhi 110002
www.koganpage.com		India

© Amy Brann, 2015

The right of Amy Brann to be identified as the author of this work has been asserted by her in accordance with the Copyright, Designs and Patents Act 1988.

ISBN   978 0 7494 7237 5
E-ISBN 978 0 7494 7236 8

British Library Cataloguing-in-Publication Data

A CIP record for this book is available from the British Library.

Library of Congress Cataloging-in-Publication Data

Brann, Amy.
   Neuroscience for coaches : how to use the latest insights for the benefit of your clients /
Amy Brann.
     pages cm
   ISBN 978-0-7494-7237-5 (paperback) – ISBN 978-0-7494-7236-8 (ebk)   1. Personal coaching.
2. Neurosciences.   3. Performance–Psychological aspects.   I. Title.
   BF637.P36B73 2014
   158.3–dc23

Typeset by Graphicraft Limited, Hong Kong
Print production managed by Jellyfish
Printed and bound in the UK by 4edge Limited

CONTENTS

PART SIX   Neuroscience of classic coaching areas 113

THANKS

This book came about because of two main groups of people. First, the coaches who are committed to serving those they work with. Their questions and requests for resources and training directed us to this project. Second, the neuroscientists committed to discovering more about the brain. The research to which they dedicate themselves all around the world enables us to also learn more about this incredible organ.

It was my pleasure to work again with Liz Gooster, the talented coach and editor who always asks direct questions that result in a better-quality manuscript in the end.

I'm also very grateful to Paul Carter, '@neuropsyguy', who not only thinks about things in a very special way but is also able to communicate them clearly and directly. His insights and suggestions meant this book was much better than it would otherwise have been.

Finally, thanks goes to Stuart, my supportive and loving husband who wasn't sure that starting to write a book with an 11-month-old baby was a good plan but made it work nonetheless. Thanks to my mum for making it easy to leave our gorgeous bundle with her – because I saw how much fun she has. Thanks to Jessie for helping to keep me sane. And thanks to Jessica who teaches me so much about the brain, love and really being alive each day.

PRELUDE

Introduction

We consider why neuroscience is becoming an expected area of knowledge for coaches and how to approach this fascinating field.

Part One – Brain areas

Here we look at various areas of the brain that are useful for coaches to be aware of. The chapters in this section are:

1 Prefrontal cortex
Affectionately thought of as the CEO or conductor of the brain responsible for higher-level cognitive functions, including attention and processing.

2 Basal ganglia
Key to storing routines, repetitive behaviours and thoughts – the home of habits.

3 Striatum
Involved in pleasure, reward, motivation, reinforcement learning, fear and impulsivity.

4 Insular cortex
Enables a degree of awareness and is involved in our ability to be 'in tune' with ourselves.

5 Amygdala
Part of the limbic system which is heavily involved in regulating our emotional life.

6 Anterior cingulate cortex
Primarily responsible for conflict or error detection.

7 Hypothalamus
Responsible for various metabolic processes and synthesizing and secreting neurohormones.

8 Hippocampus
Has roles in the consolidation of information from short- to long-term memory.

The structure of each chapter addresses:

- What is this brain area?
- Why is it important to me as a coach?
- What can I do with a client now having understood this?

Part Two – Brain chemicals

Part Two is about the brain chemicals that coaches will observe and hear about the effects of within their clients. The chapters in this section are:

9 Cortisol
Famously known as the 'stress hormone' because it is released in response to stress.

10 Dopamine
Involved in how we think and behave and our feelings of motivation, reward and attention.

11 Oxytocin
Involved in social behaviour, increasing trust, decreasing fear, increasing generosity and also cognitive functions.

12 Adrenaline
Best-known for its role in the fight-or-flight response.

13 Serotonin
Important for mood regulation, appetite, sleep, memory and learning.

14 GABA and glutamate
The king and queen of neurotransmitters.

The structure of each chapter addresses:

- What is this brain chemical?
- Why is it important to me as a coach?
- What are some scenarios where knowing about this would enable me to best serve my client?
- What can I do with a client now having understood this?

Part Three – Foundational brain concepts

Part Three looks at the foundational things that a coach needs to understand about the brain and what this means to them. These underpinnings are vital for the next three parts of the book. The chapters in this section are:

15 Neurons and synapses
16 Neuroplasticity

17 Threat/reward response

18 Neuroimaging

19 Working memory

20 HPA axis

21 Mirror neurons

The structure of each chapter addresses:

- What is this?
- Why is it important to me as a coach?

Part Four – 22. Brain networks

The approach adds a level of understanding that will take you from a basic comprehension to an intermediate one. Rather than seeing each area of the brain as independent we look at three instances where multiple regions work together in a network.

Part Five – 23. The quantum brain

Part Five offers the opportunity to explore the possibility of viewing the brain through a quantum lens – hugely controversial and, subsequently, equally exciting.

Part Six – Neuroscience of classic coaching areas

This part looks at the neuroscience behind classic coaching tools, techniques and concepts that coaches will be familiar with. Neuroscience brings credibility, precision and a new perspective to these areas. The chapters in this section are:

24 Self control/willpower

25 Habits

26 Optimism

27 Goals

28 Mindfulness

29 Flow

30 Motivation

31 Decision making

The structure of each chapter addresses:

- What is this?
- Why is it important for coaches?
- What is the underpinning neuroscience?
- What are any interesting studies in this area?
- What can I do with a client this afternoon?

Part Seven – Neuroscience of not-so-classic coaching areas

This part looks at some tools and concepts that are not frequently used by coaches or necessarily consciously addressed by them. The chapters in this section are:

The structure of each chapter addresses:

- What is this?
- Why is it important for coaches?
- What is the underpinning neuroscience?
- What are any interesting studies in this area?
- What can I do with a client this afternoon?

LIST OF ABBREVIATIONS

ACC	anterior cingulate cortex
ADHD	attention deficit hyperactivity disorder
AON	action-observation network
CAT (or CT)	computerized axial tomography
CRH	corticotropin-releasing hormone (CRH)
dACC	dorsal anterior cingulate cortex
DMN	default mode network
DMPFC	dorsomedial prefrontal cortex
dPFC	dorsal prefrontal cortex
dlPFC	dorsolateral prefrontal cortex
EEG	electroencephalography
EPR	Einstein, Podolsky and Rosen paradox
fMRI	functional MRI
FPN	frontal parietal network
GABA	gamma-aminobutyric acid
HPA	hypothalamic-pituitary-adrenal
HSAM	highly superior autobiographical memory
IA	interoceptive attention
LTP	long-term potentiation
MBI	mindfulness-based interventions
MEG	magnetoencephalography
MFN	medial frontal negativity
MRI	magnetic resonance imaging
MSG	monosodium glutamate
MTL	medial temporal lobe
NA	nucleus accumbens
NLP	neurolinguistic programming
OCD	obsessive compulsive disorder
PET	positron emission tomography
PFC	prefrontal cortex
PTSD	post-traumatic stress disorder
rIFG	right inferior frontal gyrus
SAD	seasonal affective disorder
SCT	social cognitive theory
SPECT	single photon emission computed tomography
SSRI	selective serotonin reuptake inhibitor
tDCS	transcranial direct current stimulation

vACC	ventral anterior cingulate cortex
vlPFC	ventrolateral prefrontal cortex
vPFC	ventral prefrontal cortex
vmPFC	ventromedial prefrontal cortex
VTA	ventral tegmental area
WOLF	work-related flow

Introduction

What is a coach?

'An expert in facilitating self-directed neuroplasticity.' Perhaps not the answer you normally hear, but certainly my favourite and one I use whenever I'm asked. Jeffrey Schwartz actually introduced me to this idea and I like to give credit to ideas I develop or share. Neuroplasticity is the brain's ability to change. People are incredible and have a huge amount of potential. Coaches are skilled at working with individuals to help them bring out the best in themselves and rewire their brains to make it easy for those changes to become the norm.

Going back in time, a *coche* was the word for a wagon or carriage. Essentially it was and still is a vehicle that takes a person from A to B. A human *coach* now appears in many fields. Perhaps the best-known is still the sports coach whose responsibility it is to intensively train the people they work with. In the last 30 years, however, the role and definition of a coach have developed. It is common for managers in organizations to act as coaches. It is even hoped that parents at times act as coaches to their children.

Inevitably you will have your own definition of what a coach is and what they are expected to do within that role. There are now a huge number of different models, strategies and training programmes around coaching. We believe that an understanding of neuroscience will become expected to *underpin* everything that sits on top of what a coach does. Neuroscience is not claiming to be better than anything you currently do, nor is the suggestion that this should replace what you do. Rather we suggest it will enhance your understanding of those you work with.

What is neuroscience?

Neuroscience is the scientific study of the nervous system. This discipline and its research are extensive and cover a lot, including the development of the brain, its anatomy and how it works. It looks at what happens when things go wrong, be it neurological, psychiatric or neurodevelopmental. It is a big field, and when positioning such a game-changing underpinning to coaching it is worth having a broad overview before focusing in on the parts most relevant to us. Today neuroscience has an interdisciplinary nature and liaises with disciplines such as mathematics, linguistics, engineering, computer science, chemistry, philosophy, physics, psychology and medicine. Several of these areas are of less interest to a coach.

Here are some of the areas you may hear or read about. *Affective neuroscience* is focused on how neurons behave in relation to emotions. *Behavioural neuroscience* is the study of the biological bases of behaviour, so it looks at how the brain affects behaviour. *Clinical neuroscience* is not usually of much interest to us as coaches because it focuses on disorders; however, a lot of interesting discoveries have come from this field. *Cognitive neuroscience* looks at the neural base for higher cognitive functions. *Cultural neuroscience* studies how our beliefs, values and practices are shaped by and shape the brain, minds and genes – a fascinating field that we hope will add lots to our understanding over the coming years. *Molecular neuroscience* looks at the individual molecules within the nervous system. *Neuroimaging* is the field that is expert in the various scanners that are used across the board. *Social neuroscience* brings the biological systems into the realm of social processes and behaviours. Both disciplines help refine and inform the other.

From the perspective of coaches, neuroscience is the field that can inform them on important things about the brain: things that are key to new ways that you work with clients and also that underpin things you are already familiar with. Neuroscience can explain why and how coaching works. It can enlighten coaches as to things to pay particular attention to. It can warn against other practices. As it is fundamentally concerned with the way that the brain works, its value to those working with other people's brains is quite vast.

At this point it is worth reiterating that neuroscience is not in any danger of taking over the skills, training and discipline that is coaching itself. It can certainly add to what is already there, but does not replace the great foundations that a coach has.

What can neuroscience offer the coaching world?

- An understanding of what happens when a person is being coached.
- Ways to create the best coaching environment for clients.
- For coaches who use particular models, an understanding of how they might work or an opportunity to stop using outdated concepts.
- A foundational understanding of the brain through which to evaluate everything.
- The opportunity to ask better-quality questions and consider what interventions would best serve your client.
- A focus on important considerations as to how to get desired action occurring: for example, mental stress leads to habit behaviour rather than goal-directed behaviour, so this informs how/when/what goal-directed behaviour should be attempted.

Our trained coaches tell us that it gives them a sense of credibility when they can explain the research that backs up what they are doing with their clients.

What is the vision?

The skills and knowledge that a good coach possesses can be so instrumental to a person living a good quality of life that it is my hope that in the future everyone is equipped. Imagine schools where teachers could effectively coach; organizations where managers, leaders and peers chose to coach; even, dare I suggest it, governments where coaching rather than attacking was the norm. With what we now know about the brain, and how people work, we have a huge opportunity.

People possess the precious ability to take in information, think and then act. This is grossly undervalued by individuals and by organizations. Coaching has the potential not only to help the person you are working with improve in these skills but to have a positive ripple effect.

Different ways of coaching

There are many models or ways of coaching out there. We've yet to come across one for whose practitioners we don't think neuroscience would be useful. Robert Dilts proposes that a coach has many roles at different times. He suggests that a coach may need to be in the role of guiding and caretaking, coaching, teaching, mentoring, sponsoring or awakening. Dilts elaborates on these roles in his excellent book *From Coach to Awakener* (2003). Here these roles are explored, sometimes adapted, and linked to various neuroscientific principles or areas of interest. The role of this guide or caretaker can be considered to be providing support with respect to the environment where the changes are taking place, like a guide who holds a hand to take a child or blind person down a path from one place to another. It is expected, of course, that the guide knows where they are going and how to reach that desired destination. The caretaker is tasked with providing a safe and supportive environment. Dilts suggests that in this role it is important to attend to the external environment, provide what is needed and remove distractions.

This role, like all of those set out by Dilts, will be working on multiple neurological levels, utilizing different brain areas and chemicals, and would benefit from paying attention to various things that the scientific research is pointing us towards. None of these do we know for sure without doing brain scans and taking blood samples, and even then there are things we would not know in each individual situation. So, when we talk about what is going on for people in their brains and bodies when they are being coached it is normally an informed guess. We do know that in a trusting relationship, such as between coach and client, the release of oxytocin is triggered. This will be beneficial for both parties. At a basic level, by supporting someone you might be lowering their cortisol levels and enabling them to hear more clearly. If there is a high level of trust, which, as an effective coach, you hope would come at some point if not initially, then any threat response may be decreased and the person may be able to think more effectively. By attending to the external environment and removing distractions, you are creating a better environment for the prefrontal cortex to focus and be effective and efficient in high-level information processing and decision making.

Small 'c' coaching can be considered as focusing on the behavioural level of performance. It is suggested that this resembles sports coaching and involves promoting conscious awareness of resources and abilities. Careful observation and feedback are utilized to strengthen a person's abilities. It is often context-dependent so it might be used in areas such as 'style coaching', 'business coaching' and 'wealth coaching', where the coach is suggesting new behaviours.

This behavioural level of focus is very important as it leads directly to our results. The question becomes: what can influence our behaviour? Here we are directed in neuroscience to the underpinning of habits, how we are motivated, how we make decisions and to interesting new areas such as nudging and priming.

Teaching involves new learning. It can involve increasing competencies or knowledge or strategies for thinking and doing. At times during a coaching relationship it may be important to be able to share something you think will help a person if you notice a gap in knowledge or competency. This is very obvious with neuroscience where there are things a person is doing that are not getting them the results they want, and where an acquisition of new information could really help. Learning new things and strengthening new competencies are important skills. Various brain areas are involved in these processes, for example the amygdala, hippocampus and basal ganglia.

Mentoring is described by Dilts as guiding a person to discover their unconscious competencies and overcoming internal resistance. He cites a belief in the person and validating their positive intentions as important components. The mentor has an impact on the beliefs and values of their mentee as their example unveils a little more of the person. Mentoring may involve mirror neurons or the action-observation network (covered in Chapter 21), especially when the mentor is a good example of something that the person being mentored wants to improve upon. It has been suggested that people learn tennis best by initially watching a good tennis player and then in essence copying them.

Sponsoring is described by Dilts as the process of recognizing and acknowledging the identity of the other person. This role of looking for and then helping to secure the innate potential that people possess is a critical role and one that is often forgotten in a corporate environment – with dire consequences. The idea is to connect with and validate someone deeply. This process enables the value within a person to manifest more fully. Sponsoring may be having effects at the level of the quantum brain. Things that would be more widely accepted now would be the potential of increasing an individual's serotonin and dopamine levels, activating the reward response and reducing any negative effects associated with loneliness.

The final role that Dilts proposes a coach holds is that of awakener. This is suggested to awaken others through the coach's own integrity and congruence. The idea here is to connect with missions and visions. The awakener role has an intensity about it, and is perhaps where real strengthening of new neural pathways occurs.

Why will a coach be expected to have an understanding of neuroscience?

It has been said that the gap between knowledge/theory and action is around 20 years. Since through the field of neuroscience we are now so much more equipped to understand what is going on in the brain it is only a matter of time before some people know more than others. First adopters in various spin-off fields utilizing neuroscience are enjoying the benefits of this fascinating research. Neuroeconomics is a great example of a field that was working well without the 'neuro' component. When economics turned to neuroscience to see what it could learn, additional fascinating insights emerged. This is an exciting time for the field of neuroeconomics and one we expect will become a routinely available module for those studying pure economics in the near future.

With the fast rate that we are learning about the brain and mind we have already amassed a lot of data. There is a proportion of coaches who are equipping themselves with this information. It makes sense, right? If you are going to be working with people's minds and brains then it is logical to know as much as you can about how these things actually work. You have started on this journey by reading this book. There are coaches who are also taking programmes in this area and equipping themselves with an in-depth understanding of neuroscience for coaches.

Organizations and individuals will continue to be discerning in how they choose their coaches. They will want the best, the most reliable and the most cutting edge. This means that coaches are expected to keep their knowledge up to date. When new information becomes available, especially knowledge that is so fundamental to the core of what one does as a coach, it is becoming expected that you are skilled in it.

How to approach the field

When coaches ask me this question I have several levels of response. The quickest answer is to approach the field 'curiously'. If you only focus on one thing let it be your curiosity. This is a fascinating field that will develop a great deal more over the coming years, so now is the time to cover the basics and be curious about what you are exploring.

The next level of answer is to be *confident* in what you don't yet know. Whenever I give talks on this topic, or on topics that are core to our main work of applying neuroscience to organizations, people comment that it is refreshing to work with someone who is comfortable saying that they do not know something. Rest assured that the field of neuroscience is a huge one. It would take you a long time to get to know everything in the field, and new things are discovered every week. Experts in their fields tend to be just that, expert in their field. That field is often very specific. An expert in decision making may have very little awareness of the research being done on false memory. There is no shame in not knowing everything – however, there can be in pretending you do.

There has been suggestion that some people are turning to neuroscience in a superficial way, simply because it is a hot or sexy topic at the moment. That may be the case. People fall for gimmicks – they read a headline in a newspaper or even on what they consider a credible website and then use this surface-level piece of information to 'prove' something. Taking these shortcuts is very unwise in science. Evaluating the source, going back to the research and getting a full understanding is the best option to build reliable foundations and to be credible. In fields where you are working with people's brains and minds, such as coaching, there will always be a place for understanding the science. It is illogical to suggest that neuroscience for coaches could be anything other than a positive step towards a better future.

What are the boundaries/limitations?

For coaches who are very new to the idea of neuroscience, having a contribution to make to their field in this area can be a challenge. Their expectation is that there will be specific new jazzy tools and techniques that are 'neuroscientific'. They want to pull out of their bag their neuroscience tool and have their client go through a nicely packaged exercise. This desire is understandable: we like things to be simple, straightforward and easy to position with clients. However, good and experienced coaches know that there is a lot more to coaching than simply doing exercise after exercise with a client. This is where neuroscience comes in.

In this book we have at times included a section called 'What I can do with a client this afternoon'. Sometimes there is a tightly correlated exercise that is nice and easy to complete and that would be great to explore with a client, if this is how you choose to coach. At other times there is not. It is actually even more exciting when there is more to it than a simple exercise! (This is a mindset shift for some coaches, while obvious to many others.) Building up an understanding of how your client's brains and minds work is where the true magic lies.

Take, for example, the area of habits. We know a lot about how habits are formed in the brain and what areas are involved in these processes. There are many suggestions that can be used to support those you work with to gain increased control over their lives. However, there are also recent advances in our understanding of how habits work. We know that a small area of the prefrontal cortex is also involved in habits and specifically the 'turning off' of a habit. A natural question that people ask is: 'Can we just turn off bad habits, then?' Or, perhaps: 'How do we turn off that area of the brain, then?' Both of these questions are great and valid lines of thought – however, we don't have a safe and practical answer just yet. Instead, work with habits currently involves creating new ones that become stronger and more favourable than the old ones.

The challenge this book faces

One of the challenges for a book like this is that we are discussing a field that is rapidly developing. New studies are being done every week and some of these inevitably give us information that is contrary to previous data. This book is not pretending to offer you a snazzy new comprehensive 'solve all your coaching problems' model. It is a starting point: an introduction to neuroscience for coaches.

The opportunity once the basics have been covered is to deepen your understanding of how you apply this information, by going on a well-designed programme. Here you can connect with a supportive group of coaches who can quickly accelerate your practical application of this material. You will, of course, have your own coaching style and skill sets. Suggestions are offered throughout this book only to stimulate thoughts. You need supervized practice and reflection in order to integrate what you learn into your chosen way of coaching.

There are some classic examples of what we want to avoid. Imagine a senior leader of an organization who read a book on leadership – 20 years ago – and nothing since. There may be some key concepts that still hold true, but it is fair to say that they would be a far better leader if they stayed connected to the field of leadership and updated their knowledge and skills. There is a fear with writing a book of this nature that coaches think 'Well, that's it then, I know all I need to know about neuroscience'. That simply will not be the case. We can look at this from a couple of perspectives. First, the field will be continually evolving and so it will be beneficial to keep connected. Second, this book is not *it*: while the hope is that it will serve as a good introduction, really getting the information inside you and being able to use it confidently and fluently with those you work with is the next step.

How you use this information is in the hands of each individual coach. There is a danger with being introduced to new and exciting information that we want to rush in and share it with everyone. As coaches we need to remember that we are here to facilitate self-directed neuroplasticity. Our client's ability to think is a precious one – giving information inappropriately can destroy what Nancy Klein (2002) thoughtfully terms the 'Thinking Environment'. This precious environment deserves our respect.

What is the relationship between psychology and neuroscience?

In the past, and sometimes still in the present, there has been competition between the two disciplines. They have each, at times, looked down on the other. Psychology has accused neuroscience of being too reductionist; neuroscience has said that psychology can be inefficient and messy.

The benefit of psychology is that it can highlight phenomena that can prompt further exploration for mechanisms. Neuroscience can look at the more fundamental level and discover functional capacities of neural components. Links can be made here between behaviours and brain structures and function. Currently, while the two disciplines overlap, they are often working at different levels of things: neurons versus personality.

A useful analogy that can be drawn here is that of the Formula One car. Two main parties are involved in driving this car and, hopefully, winning races: one is the driver (the psychologist) – they are in peak physical condition; they know every detail of how their car performs (or are working towards this). If they want to go faster they know what buttons they need to press. When they have tricky bends coming up they know which way they need to turn the wheel and by how much. They have normally driven a track multiple times and can predict what the car will do at any given point. The other main party is the mechanics (the neuroscientists). These guys know what is under the hood of the car. They are focused on the nuts and bolts of the engine. They know that if they adjust some fluid by even just a small amount it will have an effect. They are interested in the most fundamental components of the car. Together the driver and the mechanics make a good team. Together they can each learn from one another and try different things.

So let's open the bonnet and take a look at what's inside!

PART ONE
Brain areas

'What part of the brain does that?' – this is a common question that is natural, but often unanswerable. With all the advancements in our ability to see into the brain while it is doing things, particularly through MRI (magnetic resonance imaging) scans and EEG (electroencephalography), people frequently have the expectation that we can explain everything by linking it back to a specific area of the brain 'lighting up'. Fortunately, the brain is a little more complex than this. Despite possessing a fixed anatomy our brains exhibit diverse functionality. The question of how our brain does that is explored further in Part Four on dynamic functional connectivity. However, we first need to start with the foundations and then build up from there.

One of the most wonderful realizations felt by many of the participants of our Neuroscience for Coaches programme during the first module is that things are more layered than they previously realized. There is more going on in their client's brains than they thought. Initially some coaches hope that they would be able to identify, with certainty, an area of the brain that is responsible for a certain problem their client is dealing with – and then fix it. They quickly realize, however, that this will not be the case.

So what is the point of learning about the different brain areas? There are several answers to this question. The first is that our knowledge shapes our filters and our questions. As coaches these are two of the most valuable things we have. Consider your filters and questions before you had any coach training... and afterwards. One hopes that they are quite different. It is the same once you learn about the different areas of the brain (and everything else you cover in this book and then during your further studies).

The second point of importance for learning about the different brain areas is to give you a solid foundation. The concepts we look at in this book have clear and direct applications to coaching, so when we start talking about mindfulness and the effect it has on the prefrontal cortex then it makes sense to know what the prefrontal cortex is, what it looks like when it isn't working well, what it needs to work well – and lots more.

The third benefit is that at times you may want to discuss the areas of the brain with your clients, offering them an additional perspective to their situation. We cannot know for sure whether a particular brain area is involved with a problem or situation they are working on, but most of the time we don't need to. There are thousands of examples of how this would work in real coaching situations, in Chapter 2 on the basal ganglia we will look at just one. For a client who is working on changing their habits, being able to explain how and why the brain works with habits can be really useful to them.

It is important to bear in mind as we go through Part One that each time we get to the section 'What can I do with a client now having understood this?' what is being offered are only suggestions. What will work and be appropriate in one situation is not the same for another. You as the coach are best placed to blend the science with your art at this point and balance everything you know in order to make an informed decision about what would best serve the person you are working with.

Before we dive into the different anatomical areas of the brain it is worth taking an overview perspective and touching on something you may have already heard of: the triune brain.

What is the triune brain?

The triune brain is a model that was developed by physician and neuroscientist Paul D MacLean in the 1960s (MacLean, 1990). It describes the evolution of the brain.

The hypothesis is that there are three parts to the brain: the reptilian, the paleomammalian (better known as the limbic) and the neomammalian. The first, the reptilian, comprised the basal ganglia and it was suggested that this complex was responsible for the instinctual behaviours. So drivers like aggression and dominance were said to come from this reptilian brain. It is the oldest in evolutionary terms.

The next in evolutionary terms is the paleomammalian complex. It contains the brain areas including the amygdalae, hypothalamus, hippocampal complex and cingulate cortex. The 'limbic system' was used to refer to this group of structures that are so interconnected. MacLean suggested that this limbic system was responsible for things like motivation and emotion involved in feeding, reproductive behaviour and parental behaviour.

The newest part of the brain, according to MacLean, is the neomammalian complex. This is basically the cerebral neocortex. It is responsible for our higher cognitive functions such as planning, perceiving and language. While the model is still in favour in some circles there are some challenges. Some of the arguments against this model are that:

- Some recent behavioural studies call into question MacLean's thoughts around the reptilian brain.
- Anatomically, the basal ganglia were also found in amphibians and fish (not just reptiles).
- Birds have been documented to have sophisticated cognitive abilities and language-like abilities.

- Not only paleomammals care for their children/offspring – even some fish do it.
- Crucially, it is now suggested that the neocortex was present in the earliest mammals.

Linked to this topic is the concept of the limbic system itself. There is a lot of controversy around this term. While some still use it freely, others say there is too much wrong with it. For example, the hippocampus is a key part of the limbic system. The idea is that the limbic system is the emotional centre of the brain while the neocortex is the cognitive zone. When damage is done to the hippocampus we see that people suffer from memory problems. Neuroscientists keep changing the boundaries of the limbic system. Joseph LeDoux suggests we should abandon it altogether.

Why is the triune brain model important to me as a coach?

This model is fairly widely espoused in non-scientific circles. Having an understanding of it and what the science community are saying about it puts you in a more informed position.

Yet while this model is very attractive in its simplicity, it doesn't have the backing of neuroscientists across the board. So the suggestion here is to use with caution.

Prefrontal cortex

01

Affectionately thought of as the CEO or conductor of the brain responsible for a lot of higher-level cognitive functions, including attention and processing.

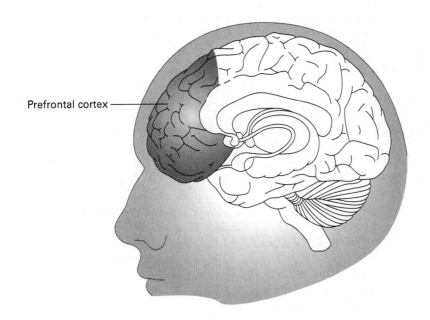

Prefrontal cortex

What is this brain area?

The prefrontal cortex (PFC) is an area of the brain found at the front of the head, behind your forehead in your frontal lobes. Evolutionarily it is one of the newest brain areas and is responsible for a lot of cognitive functions. The dorsal prefrontal cortex (dPFC) is interconnected with brain areas that are involved with attention, cognition and action. The dorsolateral prefrontal cortex (dlPFC) is known for being involved with short-term memory and is implicated in self-control. The ventral prefrontal cortex (vPFC) is interconnected with brain areas involved in emotion. The ventromedial prefrontal cortex (vmPFC) performs risk-benefit analyses after receiving inputs from the amygdala and other parts of the frontal lobes.

There is a famous story in the neuroscience world that involves a man called Phineas Gage. In 1848 Gage was a 25-year-old construction worker. He had been a respected friend to the men he worked with and was good at his job. One day there was an explosion while Gage was tamping powder with a fuse in a hole (before sand had been poured in). The rod weighed almost a stone, was 3 feet 7 inches in length and 1.25 inches in diameter. As it flew into the air it pierced Gage's left cheek, going through the base of the skull, traversing the front of his brain and exiting at high speed out of the top of his head. The rod mostly destroyed his frontal lobe, including his prefrontal cortex.

The first thing that amazed bystanders was that Gage was able to walk, talk and be 'normal' (remember that part of his brain was now lying some distance away on the end of a rod). The second thing is that he survived the predictable infections that in 1848 were treated without antibiotics. Yet although he was physically intact, his previous likeable personality was not. He became 'fitful, irreverent, indulging at times in the grossest profanity which was not previously his custom, manifesting but little deference for his fellows, impatient of restraint or advice when it conflicts with his desires, at times pertinacious obstinate, yet capricious and vacillating, devising many plans of future operation, which are no sooner arranged than they are abandoned... A child in his intellectual capacity and manifestations, he has the animal passions of a strong man.' The damage to his brain – as we can see (thanks to the skull being dug up, examined, and scanned) – was in the prefrontal cortices. His personality was changed forever and he was unable to make good choices.

Some more recent studies on people with prefrontal cortex injuries have developed our understanding of this region of the brain. When individuals were asked what an appropriate social response would be under given circumstances they would give an appropriate answer. However, when they were actually making choices in real time they would select behaviours that were immediately gratifying even though they knew in the long term it wasn't the best plan. Our ability to delay gratification is very important to humans and is an example of a healthily functioning brain. Modern examples of short-term gratification at the expense of long-term benefits would be such as eating too much fatty foods, buying things today rather than saving for to-morrow, having extra-marital affairs and sitting down with a glass of wine in front of the TV rather than hitting the gym.

Why is it important to me as a coach?

The PFC is involved in:

- all our executive functions;
- our ability to plan;
- decision making;
- expressing our personality;
- aligning our thoughts and actions with internal goals;
- moderating social behaviour.

The psychological term 'executive functions' covers a wide range of activities. It includes the ability to determine good and bad, same and different, process-ing the future consequences of current activities, working towards defined goals, suppressing socially undesirable urges and predicting outcomes.

A professor of mathematical psychology, David Meyer, performed a wonderful experiment. He invited a group of young adults to test what happens when people are switching between things quickly. The experiment involved the participants working out mathematical problems and identify-ing shapes. When they had to switch between the tasks their accuracy and speed decreased as compared to when they could perform one task and then the other. In some cases multitasking added 50 per cent to the time required. Imagine a person working a 12-hour day and achieving the same results as a person working 8 hours, but with more mistakes and less elegance simply because the 12-hour day person was multitasking. 'Not only the speed of performance, the accuracy of performance, but what I call the fluency of performance, the gracefulness of their performance, was negatively influenced by the overload of multitasking' (Meyer *et al*, 2001).

Under stress our prefrontal cortex does not function well. The neural circuitry and neurochemistry of the PFC can be changed by our experiences. Stress during childhood and adolescence can be particularly impactful on the structure and function of the PFC, although these effects are not neces-sarily permanent. In adults it has been shown that even mild acute (as opposed to chronic) stress can have a rapid and dramatic impact on our PFC's ability to function. It can affect creativity, flexible problem solving, working memory and other processes. Many stressed-out people find it quite a wake-up call to discover that chronic stress literally can cause the architecture of their PFC to change. One study looked at the effects of psychosocial stress after one month. It showed that the subjects had impaired attention control and disrupted functional connectivity within a frontoparietal network that mediates attention shifts. The good news was that after one month of re-duced stress their responses became normal again.

In a scenario where someone feels overwhelmed, with too much work to do or with a goal that is too big, then it can be useful to break it down into what needs to happen today. For example, if a budget meeting takes place and results in a huge financial goal that needs to be met, it could help to break it down into a weekly target. This could re-engage the dorsolateral PFC (short-term memory) by chunking things down into small short-term bite-sized pieces.

Conversely, when someone is feeling very anxious it may be that their ventromedial PFC is activated. Consider a personal coaching scenario where a person's spouse tells them that they no longer feel loved by them. The person may feel an immediate threat, imagining what would happen if they broke up and how difficult things would be in lots of ways. An aim here would be to reduce the immediate threat. Change the time frame from a very short-term one to a long-term perspective. It also may be useful to shift from an empathetic state to a cognitive one. Get planning – what could be done over the next month to help move the situation in the direction your client wants it to go?

What can I do with a client now having understood this?

In some coaching situations it might be appropriate to step into the teacher role and share some fundamental things about the PFC. This would be especially relevant to anyone working on their productivity, their efficiency or effectiveness.

What do you need to know about the prefrontal cortex?

- Its role is classically described as like that of a CEO (if you are business-minded) or a conductor (if you are musically minded) – in short, it is the boss, it is responsible for your 'executive' functions, which means your ability to think, choose, plan, etc.

- Over the years it has developed considerably, and recent studies indicate that meditation increases its size further still (see Chapter 28 on mindfulness).

- It is hugely energy-hungry, but gets drained quickly.

- Stress impairs its ability to use energy.

- Structurally, it is part of your frontal lobe, which is the area at the frontmost part of your brain.

So what is the motivation for ensuring our PFC is in good working condition? When our PFC is not working optimally we find ourselves:

- feeling lazy;
- feeling lethargic;
- uninspired;
- easily distracted;
- being poor at completing things;
- fixing attention on repetitive negative thoughts;
- being disorganized;
- being forgetful;
- feeling overly emotional.

On the other hand, when the PFC is working in tip-top condition you can look forward to:

- intentional awareness;
- a long attention span;
- being able to contemplate possibilities;
- being able to plan;
- being able to stick to the plan;
- focusing easily.

When the PFC is overworked is doesn't function well at all. To rectify this situation, check out the recommendations below. If the PFC is not functioning properly this makes it very difficult to be effective, let alone efficient, which is often very frustrating when we are trying to get through the day. At such times it is quite common to feel that something is wrong and to perhaps fall back on old habits. This is a form of survival mechanism. When people find themselves micromanaging, controlling or punishing there is often a brain deficit involved. This doesn't always mean that something is overactive; an underactive PFC causes problems too. Dopamine, you'll soon find out, is the neurotransmitter that has many functions for the brain including reward, motivation, working memory and attention. When the brain has a lowered ability to use or to access dopamine it means that other brain areas aren't being quietened down in order to enable the brain to focus on one thing at a time. This makes life difficult for us because we struggle to focus, so become less efficient.

As a coach you could also step into a mentoring role and lead by example, by following these recommendations:

- Turn off the e-mail function on your mobile phone in the evening so that your brain has downtime before you start work the next morning.

- Prioritize the big weekly tasks first then smaller tasks on a daily basis (experiment with the night before or first thing in the morning to see what works best for you).
- Turn on e-mails only at certain times of the day.
- Monotask for short- and long-term benefits.
- Pick small things you know you can achieve and then do them – this could raise your dopamine levels.

Equally, you could work with your client to help them experience their prefrontal cortex working well. Take them out of their normal working and thinking environment – for example, if they are normally office bound, take them into the countryside. Help them to connect with their senses, feeling the wind in their hair, the sun (hopefully) on their skin and breathing in all the different smells. The hope would be that this change of environment would invite some fresh thinking.

Basal ganglia

02

Key to storing routines, repetitive behaviours and thoughts – the home of habits.

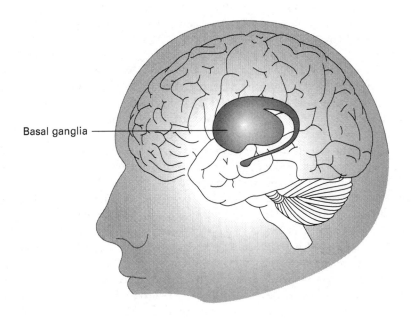

Basal ganglia

What is this brain area?

A more accurate question is 'what are they?' There are billions of basal ganglia and they are key to storing routines, repetitive behaviours and thoughts. The patterns that your daily life is filled with are recognized, stored and repeated by your basal ganglia. The basal nuclei (another term for the basal ganglia) are in fact a group of nuclei, made up of several components, that act as one unit. You'll find them at the base of the forebrain. Along with habits, the basal ganglia are associated with cognitive and emotional functions. The main components of the basal ganglia – you do not need to remember this, it is included only because some people enjoy a high level of detail... I rarely refer to these components in practice – comprise the striatum (caudate nucleus and putamen), the globus pallidus, the substantia nigra, the nucleus accumbens and the subthalamic nucleus. You will hear lots more about the striatum and the nucleus accumbens in the following chapter and these will start to stick in your brain – so continue to breathe, relax and enjoy exploring neuroscience.

The basal ganglia works using 'if-then' coding and has lots of mini programmes, or maps, stored. Every habitual thing you do has a map. For example, if when you turn on your phone in the morning you immediately check LinkedIn to see what is happening, then your basal ganglia is coding 'if the phone is turned on in the morning then check LinkedIn'. The instructions are present for exactly how to do this. This brain region is very well connected. Information from most other brain areas is sent here.

A study that looked at the recognition of patterns in the basal ganglia when the subject is unconsciously aware of a pattern being present was carried out in Montreal. The volunteers were split into two groups. Both groups had to press one of four buttons on a keyboard to correspond with a light flashing in a certain location on a screen. One group watched the light appear in a random order while the other group were given a pattern. This pattern was difficult to consciously detect most of the time, but their basal ganglia could be seen via the brain scanner to be picking it up. It has been said that it can take as few as three repetitions for the basal ganglia to pick up a pattern. At times, individuals were able to identify the pattern and they could type their sequence 30–50 per cent faster than if there wasn't a pattern in the sequence. This underpins how a large proportion of our day is run on autopilot.

The basal ganglia are influenced by signals coming from other parts of the brain. This is important to be aware of for a fuller understanding of the brain because there is often a balance of inputs occurring. A classic example is emotional and rational inputs into a decision-making process. When we switch from one behaviour to another, which involves the activation of the motor system, the signals from other parts of the brain are key. For example, the prefrontal cortex may convey a signal telling the basal ganglia to enable a specific action to occur. Therefore, understanding how the prefrontal cortex works, and the other brain areas, is also important to fully understand how the basal ganglia will respond.

The neurotransmitter dopamine plays an important role in the basal ganglia. A neurotransmitter is a chemical that communicates a signal across the gap (called a synapse) between two neurons. There is a lot of evidence that the basal ganglia have a role in motivation, especially the limbic part of the basal ganglia, the nucleus accumbens (NA), ventral pallidum and ventral tegmental area (VTA). The dopamine neuron projections from the VTA to the NA is important in the reward system of the brain. Things that we find primarily rewarding in life such as food, sex and addictive drugs all trigger the VTA dopamine system.

Why is it important to me as a coach?

The basal ganglia are fundamental to habits and to general automatic responses. People you work with are likely to be experiencing a large part of their day on autopilot. The thoughts they have will be the same as yesterday's thoughts. The actions they take will be similar to yesterday's actions. The interactions they have with family and colleagues are likely to be quite repetitive. This is all desirable from the efficient brain's perspective. However, when the content of these repetitions is not getting the results the person wants, then things need to change. It can be useful for clients to understand how their brain is wired and how they can change that wiring. There are great benefits to our basal ganglia working in the way that it does and empowering clients to know how to work with their brains rather than against them. We want our brains to use precious energy for processing novel information and to think, but for this to be the case we need habits that support us. We explore these in depth in Chapter 25 on habits.

What can I do with a client now having understood this?

As previously acknowledged, coaching styles vary widely and many of these different approaches lead to the facilitation of self-directed neuroplasticity. One thing you could consider doing with clients involves the sharing of some of this information. Perhaps at a time when you are focusing on habits you could introduce the basal ganglia – by sharing the positive intention of this part of the brain, and how we rely on maps and circuits for things so that we do not have to think about them every time we do them. You should also share the potential downside to this, being that when we want to do something differently it can be hard. Our brain wants to be efficient and free up vital resources to be used for things that are new and require attention. Yet by seeing habits as things that serve us we can start to develop some great foundations.

Considering next coaching in the personal arena, talk to your client about their 'if-then' codes, open the dialogue and see where it takes you. It may be that a whole session is required on addressing the 'if-then' coding that is and isn't working for your client. Let us take the example of someone working on a weight-loss goal. If they see a piece of chocolate cake and then they eat it this may not be ideal. If they get home from work and then see the sofa and fall exhaustedly onto it, rather than turning around and heading out to the gym, then changes also may be a good move. If they say to themselves regularly 'I deserve a treat' and then have an Indian takeaway, that may not be ideal. These are all if-then plans coded by the basal ganglia. Simply by talking about these if-then plans and verbalizing more appropriate ones means it is more likely that the latter will be enacted. How long it would take to make new habits depends on a lot of variables, so I prefer to focus on putting the building blocks in place that will give people the best chance of turning to these new habits when they need them rather than to say 'in 30 days you'll have broken old habits and replaced them with new ones' (as this just is not necessarily the case).

In a more corporate environment our attention is drawn to the role of dopamine in the basal ganglia. Dopamine helps us to feel motivated, which is obviously very important at work. Lots of energy, time and money gets invested by organizations into trying to motivate their employees. As a manager in the role of a coach or a leader you have an opportunity to educate or mentor those you work with. At a very basic level, living a healthy lifestyle is important for brain health as well as the health of every other organ. People could ensure that they are getting enough dietary tyrosine, which is a basic building block of dopamine and is found in avocados, bananas, meat, sesame seeds and almonds. Other foundational things to consider, and whether or not these are in place, are people eating enough antioxidants from green and colourful vegetables and fruits. Antioxidants can reduce free radical damage to brain cells, including the cells that produce dopamine.

If you work in an organization, could your team do a healthy living month? This could, using some gamification principles and behaviour-change concepts, improve the choices people are making and ultimately lead to a healthier, more robust brain. Experientially helping people to understand how their basal ganglia work is another option that is open to you. Gamification is a relatively new process that is capitalized on by technology companies. Some principles that could be mapped across into organizations – this would need testing to see what concepts work best – include:

- Create a public 'scoreboard' for people to mark on it their intended workouts for that week and their actual workouts, and have stars given each week to people who do what they said they would do.

- Create a 'buddy system' for people to hold one other individual accountable to their daily number of steps targets.

- Have a bowl of fruit in the office with a set of smiley-face stickers next to it, then each time a team member eats a piece of fruit they should put a sticker next to their name on their personal scorecard – their aim is to get to two per day within four weeks.

Finally, since the basal ganglia are influenced by the prefrontal cortex (PFC) it is worth teaching your client about how the PFC functions well – in order to help them get the best from themselves overall.

Striatum and nucleus accumbens

03

Involved in pleasure, reward, motivation, reinforcement learning, fear and impulsivity.

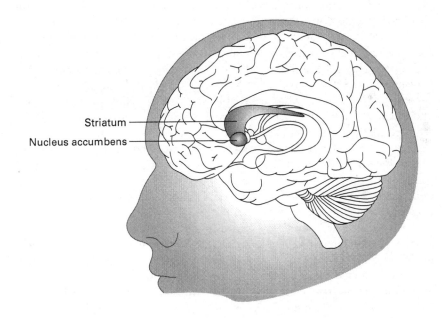

Striatum

Nucleus accumbens

What is this brain area?

The striatum is the core component of the basal ganglia. The striatum receives input from the cerebral cortex and gives input to the basal ganglia, so it acts as a sort of relay station. Why cover it separately from the basal ganglia? Because there is so much to cover and a lot of research mentions the specific components.

Although at this stage of the book there are a lot of anatomical terms being introduced, they will become more familiar as you read them again in other contexts! The ventral striatum is the part of the striatum closest to the face and consists of the nucleus accumbens and the olfactory tubercle. The dorsal striatum is the largest part of the basal ganglia and is positioned towards the back of the head (ventral always means towards the front and dorsal towards the back of the head) – it includes the caudate nucleus and putamen.

The nucleus accumbens is part of the ventral striatum, which you'll remember is part of the basal ganglia. As is often the case we have two, one in each hemisphere. Technically you can divide the nucleus accumbens into the core and the shell, with different functions, but we'll refer only to the nucleus accumbens (NA) as a whole. We know that the NA is involved in pleasure, reward, motivation, reinforcement learning, fear, addiction, impulsivity and the placebo effect.

The NA is well known for its role in addiction. Functional imaging studies show that environmental cues that are linked to people taking addictive drugs trigger the release of dopamine in the NA. For most coaches, what is of more interest is its role in processing rewards and novelty in memory encoding. It is activated when we see pleasant, emotionally arousing pictures and during visualizations of similar such scenes. It is also involved in the regulation of emotions induced by music.

One of the old but classic studies was performed in the 1950s (Olds, 1956). Electrodes were put into the septal area of a rat and the rat could push a lever to stimulate the electrode, by so doing stimulating that area of its brain. They found that the rat would continue to press the level, preferring it to even stopping to eat or drink. The conclusion was drawn that the area is the 'pleasure centre' of the brain and is involved in reinforcement learning. In human experiments the level of dopamine in people's NAs went up when they believed they were receiving money, for example, or when heterosexual men were shown pictures of attractive women. The NA is also important in the feeling of 'wanting'. We get a flood of dopamine in the NA when we are expecting or anticipating a reward. This facilitates lots of different approach-oriented and goal-oriented behaviours.

Axmacher et al (2010) looked at how we process unexpected events. It was known that we are more likely to remember unpredicted events than the predictable and routine ones. The relationship between this novelty processing and the hippocampus and NA was explored by Axmacher et al.

Prior to this particular study it was known that the hippocampus is involved in memory formation and the NA is involved in processing rewards and novel information. It was thought that the information transfer between the two brain areas was greater for unexpected events.

The study was an unusual one and the data collected had some limitations but the researchers found that the unexpected stimuli enhanced an early and late electrical potential in the hippocampus. The late signal was linked to the memory for the unexpected stimulus. In the NA there was only a late signal and it was larger for the unexpected stimulus. It appears that the two brain areas work as a team, as stated by the study leader Dr Axmacher: 'Our findings support the idea that hippocampal activity may initially signal the occurrence of an unexpected event and that the nucleus accumbens may influence subsequent processing which serves to promote memory encoding' (Axmacher *et al*, 2010).

As a coach you are often on the lookout for new, effective ways to help your clients get the most out of themselves. We know that there are strong links between behaviour and the architecture of the brain. For your clients, as designers of their futures this is important for them to be acutely aware of too.

A fascinating study that is described as the first to correlate social media use with brain activity looked at reward activity in the brain and Facebook (Fareri and Delgado, 2014). The researchers found that they could predict the intensity of a person's Facebook use by the activity in their NA. Facebook has a feature where people can 'like' information posted on it. This can be thought of as positive social feedback related to one's reputation. Each participant completed the Facebook Intensity Scale disclosing how many friends they had, how long they spent on Facebook and some general thoughts. Next the participants were interviewed on video and then their brain activity was recorded in various situations. For example, one scenario was that the participant was told that their video interview had been shown to another person and that person thought highly of them. They were also told that another person had received positive feedback. The NA was activated more strongly when the participants received positive feedback about themselves than when they heard that another had received it. This was fairly predictable. What was really interesting was that the strength of this difference corresponded to participants' reported intensity of Facebook use. Another situation the participants were put in was to perform a card task to win money. The NA response to the monetary reward did not predict Facebook use.

As is often the way with studies like this, we do not know from the results whether positive social feedback drives people to interact more on social media or if the sustained social media usage changes the way this feedback is processed by the brain.

Why is it important to me as a coach?

The striatum is very important in motivation, reward and salience. It co-ordinates motivation with movement in both lower- and higher-level functions. For example, it is involved in our ability to inhibit undesirable behaviours in a social interaction. The ventral striatum is involved in rewards. How we are motivated to do things, the value we attach to things and what we are prepared to do in relation to those things all come into play here. The caudate nucleus is an important part of the brain's learning and memory system. The putamen is also involved in learning.

Neuroimaging studies with functional MRI (fMRI) shows that the ventral striatum responds to reward. The value reward of the reward is coded for taking into account the quantity and probability of this reward. The striatum additionally is activated by novel, unexpected or intense things. The common denominator is thought to be salience (the salience of something is the quality by which it stands out relative to its neighbours). It is believed the striatum may also be involved in cognitive processes involving executive function, such as working memory.

The caudate nucleus is innervated by lots of dopamine neurons, which links with its roles in learning and memory. The dopamine system is known to be involved in feedback processing. When a person is receiving feedback we can see that the caudate nucleus is activated. The putamen affects reinforcement learning and implicit learning. Reinforcement learning is obviously very important to coaches supporting clients to get the most out of themselves and to respond optimally to their environments.

One study showed that delayed rewards lose their value for economic decisions and subsequently constitute weaker reinforcers for learning. This study aimed to use short delays to measure fMRI responses in the ventral striatum, a key reward centre of the brain. Other studies based on the dual-process discounting concept show stronger activations for immediate rather than delayed rewards.

Another study looked at the differences that the magnitude of a reward had on people. The experiment involved showing different instruction pictures to monkeys where each image represents a different quantity of fruit juice. The monkeys had to reach to the spatial location where an instruction picture was presented and they would then receive that amount of liquid. The study found that around half of the striatal neurons showed differing response levels depending on the amount of juice they would be receiving. The takeaways from this are that the striatum uses expectations that give precise information about the upcoming level of reward in order to direct general behaviour.

What can I do with a client now having understood this?

The links here to practical applications are not strong ones as the experimental data is mostly from precise research into the striatum, rather than the bigger-picture perspective that would give direct suggestions of actions. However, it would be fair to suggest that experimenting with using social rewards in a work environment could be useful. Being in the role of a coach in an environment where you manage people or are part of a team gives you a great opportunity. Recognizing people for specific great things they have done has been shown to trigger the reward response in individuals. If a team is working on a big project or task it can be a great idea to break it down into small achievable chunks. Rewarding the achievement of each chunk can help to keep everyone motivated as they progress.

You could dedicate a whole session with a client to looking at the feedback loops that are set up in their life. For example, how do people who work on their own keep motivated? (One way is through their work with a coach who can give them feedback, praise and encouragement.) At work does your client feel they have systems and relationships in place that give them quick feedback? What about in personal relationships, how do they know they love their partner in ways that make them feel loved?

Purely using your coaching skills and knowledge you can help a client to re-evaluate and set up new feedback loops that serve them better. This can lead to a more frequent and useful release of dopamine and reinforcement learning occurring.

You quite possibly already do some things that activate the NA. Whether you work in a corporate environment or elsewhere the power of visualizations is strong. When we visualize something there are real things happening in our brain. Sometimes this frame can help people who think that visualizing is a very fluffy practice. Imagining something pleasing and emotionally powerful to us activates the NA.

A manager in the role of a coach could encourage the person or group they are working with to imagine a scene when a goal has been achieved. What this is definitely not saying is that by imagining something it will magically occur. That is not what we are talking about here. We are focusing on the changes that occur internally within a person when visualizing something that makes their internal environment a more favourable one. People are more likely to be motivated and take goal-oriented behaviours with a boost of dopamine in the NA, and so visualization may make a particular outcome more likely.

In a personal coaching situation it might be beneficial to explore how often positive memories are being intentionally created between the two people involved. One way to increase frequency could be to plan unusual, different activities. For some people this could be an activity such as taking a bike ride together, going on a cooking course or going camping. The idea is to do something novel and enjoyable together. We tend to remember these kinds of things much more easily than the repetitive mundane stuff. The same principle could be applied to a corporate environment. Team away-days, bonding activities and things that take people out of the familiar office environment can be great experiences. For any coach, varying how you do your coaching sessions could sometimes be to your benefit. Perhaps you could take a walk with your client while discussing something for which movement could be metaphorically useful.

Insular cortex

Enables us a degree of awareness and is involved in our ability to be 'in tune' with ourselves.

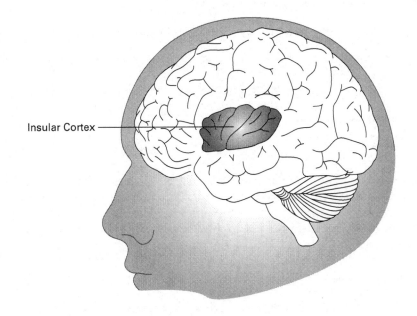

Insular Cortex

What is this brain area?

There is an insula in each of the hemispheres of the brain. They are areas of the cortex folded deep down between the temporal, parietal and frontal lobes.

The insular cortex is very well connected to the amygdala (the area important in emotional processes). The structure of the right anterior insula enables us a degree of awareness, from feeling heat to empathizing with others. Increased grey matter in the right anterior insula correlates with an increased accuracy in the ability to be 'in tune' with your inner body. The insulae are also thought to be involved in consciousness.

Why is it important to me as a coach?

The insulae are involved in a range of things including:

- emotion;
- perception;
- self-awareness;
- decision making;
- cognitive functioning;
- interpersonal experience.

Antonio Damasio proposes that the insula plays an important role in his 'somatic marker hypothesis' (Damasio, 2008). This concept is much easier to understand than it sounds. It promotes the idea that rational thinking cannot be separated from feelings and emotions. The idea is that we use bodily signals to help us make decisions. Damasio helped put the insulae on the map (as it were) and drive more research into it by suggesting that the insula plays a vital role in processing these sensations so we could use them to help make decisions.

It is proposed that subjective feeling experienced by people involves representation of bodily responses elicited by emotional events. Different people have different levels of sensitivity to internal bodily responses. In one study, people were asked to judge the timing of their heart beating. Through functional magnetic resonance imaging (fMRI) the researchers saw increased activity in the insulae and a couple of other brain regions. The level of activity in the right anterior insula predicted how accurate the individual was in judging their heart rate. Additionally this study showed that the volume of grey matter in this area correlated with both interoceptive accuracy (our ability to be aware of what is going on inside us – for example, how often our heart is beating) and subjective ratings of visceral awareness. We look at this in more depth in Chapter 28 on mindfulness.

A Japanese study addressed the effect of looking at images of painful events and images that evoked fear and rest (Ogino *et al*, 2007). Participants were told to imagine pain in their bodies when they saw the pain images. It was found that, when people imagine pain, areas of the brain involved in pain are activated, including the right anterior insula. It has been proposed that people who are prone to anxiety show an altered interoceptive prediction signal resulting in a prospective aversive body state triggering an increase in anxiety, including worrisome thoughts and avoidance behaviour. The anterior insula plays a key role in this process. Sufferers of irritable bowel syndrome have been shown to have abnormal processing of visceral pain in the insular cortex and this is related to dysfunctional inhibition of pain in the brain.

Another interesting study showed that hypnosis affected the response of the insulae during exercise. Healthy individuals were hypnotized during constant-load exercise to believe they were cycling normally, downhill and uphill. The blood flow to their left insular cortex decreased when they thought they were cycling downhill, and increased in their right insular cortex when going uphill. This and other imaging studies have shown that the insular cortex may play a big role in cardiovascular regulation. It is possible that it might modulate the physical responses based on our perceived levels of effort.

What can I do with a client now having understood this?

Some of the direct opportunities you have with clients are linked to utilizing the brain's plasticity and improving the functioning of the insular cortex area of the brain. An MRI scanner study showed that people who meditate have significantly thicker right anterior insula; mindfulness meditation has also resulted in positive scanning results showing an improvement in cortical structure. Mindfulness meditation involves a specific non-judgemental awareness of present-moment stimuli without cognitive elaboration. The fact that the brain's structure can change is very encouraging to coaches seeking to help people change and develop. For clients who are wishing to gain the benefits derived from an increased interoceptive awareness of body states then working on their insular cortex could be a good move. This can be done through various forms of mindfulness training. Mindfulness is covered in more depth in Part Six.

Some research has pointed towards yoga and Pilates as also being good for increasing insula activation. Perhaps in a corporate environment a class could be arranged on site after work once a week. Internal coaches and managers in coaching roles could lead by example and attend. In personal coaching, perhaps you could invite clients to experiment for one or two months to see for themselves whether they experience enough knock-on benefits to merit the time investment.

The benefits of increased self-awareness are well documented in fields such as organizational leadership, management and emotional intelligence. If coaching is happening as part of a 360-degree review then the opportunity to explore this is obvious. What would be different if you had such a level of self-awareness that you knew clearly how others perceived you? The neuroscientist Vilayanur Ramachandran suggests that mirror neurons may be part of the neural basis of self-awareness. We know for sure that self-awareness has benefits and subsequently, as a coach, your role can be in part helping people to increase theirs.

Amygdala

05

Known for being part of the limbic system that is heavily involved in regulating our emotional life.

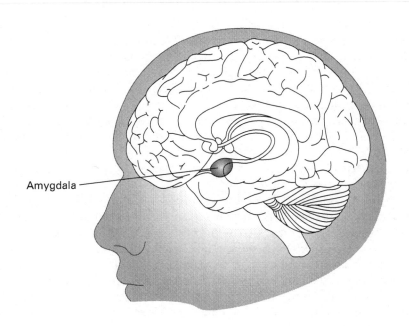

Amygdala

What is this brain area?

There are two amygdalae and these almond-shaped groups of nuclei can be found within the medial temporal lobes. They are known as part of the limbic system (the slightly controversial group of brain areas that are heavily involved in our emotional life – some neuroscientists suggest that we abandon the concept of a functionally unified limbic system, as the under-pinnings are no longer accepted as accurate). The amygdalae are responsible for different aspects of perceiving, learning and regulating emotions.

You may have heard that, according to the triune brain theory, the limbic system (including the amygdalae) developed to manage the fight-or-flight response necessary for survival. Evolutionarily it is newer than the reptilian brain (including structures such as the brain stem) but not as new as the new mammalian brain (including the cortex).

The amygdala responds to environmental stimuli that may be picked up consciously or unconsciously. It is specifically involved in motivationally relevant stimuli such as fear and reward. When we are anxious our amygdala gets activated. However, we must remember that the amygdala detects all emotions. It processes them in order of significance, and fear is very significant. It is connected to the cognitive parts of the brain, specifically the prefrontal cortex and anterior cingulate cortex (ACC). So strong are the links that when the amygdala becomes unsettled for any reason often so does the ACC. When the amygdala is activated through anxiety, the knock-on effect is that areas of the prefrontal cortex and ACC become disrupted, meaning that short-term memory, attention and our ability to make risk-benefit assessments are all affected.

There are also strong links between the amygdala and the hippocampus; the latter gets stimulated to remember details of situations and experiences. You may have also heard about the amygdala in relation to the adaptive fear response. This essential behavioural adaptation that keeps us safe is fairly well written about. If we see a file on the desk of our boss with our name on it, or what we thought was our name on it, this sensory information is quickly passed to our thalamus. The thalamus doesn't know for sure whether this information means we are in danger or not, but forwards the inform-ation to the amygdala. The amygdala takes action to protect us and tells the hypothalamus to initiate the fight-or-flight response. We realize that our heart rate and breathing rate have quickened; the slower, but equally impor-tant process that is also initiated when we see the file involves the thalamus sending the sensory information to the sensory cortex, where its meaning is interpreted. The sensory cortex identifies that there are multiple inter-pretations of the data and sends it to the hippocampus. Here, previously encountered stimuli and scenarios are taken into account. We consider whether we really saw our name, whether it could have been a harmless file, or even that it was a file we gave to our boss ourselves. The hippocampus concludes that there is no danger and sends a message to the amygdalae to tell the hypothalamus to shut off the fight-or-flight response.

However, until recently we didn't know the neuronal circuits that were being used. Sometimes a fear response is innate, other times it has been conditioned. Whenever we are, in effect, learning to fear something, the amygdalae are thought to be involved. The study that led to the identification of the neuronal circuits used mice. These mice first went through a simple behavioural task that conditioned the fear response to a sound. Once this was set up, and a fear response was observed, the researchers used advanced pharmacological and optogenetic techniques (enabling the monitoring of individual neurons). They showed that both the medial and central nuclei of the central amygdala were involved in either learning or behavioural manifestation of the fear responses. By manipulating different parts of the amygdala they could identify the subdivisions that seem to be responsible for learning to be afraid and acting afraid.

This study was even able to identify the neurons within the structures that were responsible for the components of the fear response. The relevance of this is huge. People who suffer from debilitating fear responses, such as many of the anxiety disorders or post-traumatic stress disorder, could all benefit. Further research is needed into how the selective manipulation of these neuronal circuits can be achieved most efficiently. Any coaching tools that claim to help people overcome an undesirable fear response would likely need to work on these circuits.

It is exciting to see coaches realize just how interconnected our experiences are with shaping our brains. Anxiety is a common and, to a degree, healthy response to stress. In short bursts it helps us process and cope. When anxiety continues over a prolonged period of time, however, problems can arise. The link between what is happening in our lives and what is happing in our brains is fascinating. Studies that have looked at adults with anxiety disorders have shown that they have enlarged and highly connected amygdalae. Their brains have changed in response to their experiences. Studies with lab animals that were put in environments that caused chronic stress have showed that these animals' amygdalae grew additional synapses. They also increased their synaptic connectivity.

Another study focused on children who suffered from anxiety (the specific levels were lower than what is considered clinical anxiety, however). They found that four functional neocortical systems in the basolateral amygdala were affected. The systems that deal with perception, attention and vigilance, reward and motivation, and detection of salient emotional stimuli and regulation of emotional responses were all affected. This was seen in children aged 7 to 9. The changes in these children's brains could predispose them to anxiety disorders later in life. Thus it is important to keep our focus – as coaches – on the fact that the brain can change.

The amygdala has been shown to be important in a rich and varied social life within humans. Previous research has shown that primates who live in larger social groups have larger amygdalae. In 2010 research also revealed that the volume of the amygdala positively correlated with both the size and complexity of social networks in adult humans. This holds true for both men and women, both younger and older. No such correlations were found between other structures in the brain. The suggestion is that evolutionarily the amygdalae could have evolved, at least partially, to deal with our increasingly complex social life. This 'social brain hypothesis' is a popular one. This, along with other studies, leads us to believe that the amygdalae play an important role in social interactions.

Why is it important to me as a coach?

The amygdala is important in:

- emotional responses and these drive a lot of our behaviour;
- memory;
- attention – the ability to focus in on something while excluding others;
- social processing.

Each of these things are of regular importance to coaches, as you may already recognize, and we'll see more as we progress through this book. Coaches who understand that actual changes take place in the brain as a consequence of experience are in a better position than those who do not. This fundamental point has the power to change how you approach your work with clients. Expecting change to occur quickly, even once a stimulus has been removed, is unwise. The brain needs time to rewire again.

In a corporate environment it can be the case from time to time that people get scared about something. Perhaps a reshuffle is announced and people are scared about whether their role will be changed or removed altogether. Perhaps end-of-year figures come in and someone knows their department won't have performed as well as last year. Perhaps it is as simple as being requested to attend a meeting with the boss. As a manager or leader in a coaching role it is useful to remember that it will be hard for an individual to process much else until their fears are under control. It is a very powerful emotion and can overpower other emotions and processes.

Optimism can replace fear. This is great news and often can be used to our advantage. Diverting a focus from what may happen that is negative, such as some form of telling off from a boss, to what is realistically optimistic – such as an opportunity to learn, share insights and move forward – can be useful. It is worth consulting Chapter 26 on optimism to ensure that any actions along these lines are in the best overall interests of the people we work with.

Enhancing trust can also decrease amygdala activation. Creating opportunities as a coach to increase perceived trustworthiness and trusting your client is always a valuable investment. Coaching those you work with to do the same with others is also a great plan. They may not be aware that to get the best out of someone whose amygdalae are highly activated one of their options is to build trust with that person.

What can I do with a client now having understood this?

- Be alert to a client letting fear hold them back. Question whether their amygdalae could be overreacting. Often even by posing this question you can open a dialogue with a lower fear response in place.
- Although we don't know of a specific study proving this, it is possible that just by being around fear-inducing stuff your unconscious will pick up on this. Therefore, consider experimenting with reducing these inputs, eg turn off the news, don't read a newspaper, avoid negative people.
- Deal with things – of course, this sounds obvious to a coach! Helping your client look at challenges and things that scare them is hugely important because it frees up other areas of their brain to work effectively and efficiently.
- Help your client to explore unconscious fear triggers.
- Don't allow your client to make big decisions when they are anxious, because they are unlikely to be considering the whole picture.
- Despite what people say, working under pressure doesn't create optimal brain environments. Encourage experimenting with alternatives.

Anterior cingulate cortex

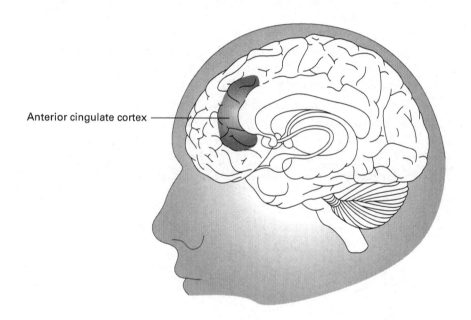

Primarily responsible for conflict or error detection.

Anterior cingulate cortex

06

What is this brain area?

The anterior cingulate cortex (ACC) can be found looking like a collar around the frontal part of the corpus callosum (the bundle of neural fibres that connect the left and right hemispheres). The primary role of the ACC is in error detection and self-correction. It deals with conflicts between autonomous and cognitive processes. The classic activity that highlights the involvement of the ACC is called the 'Stroop task'. This activity involves a person naming the *colour* of a word. The tricky bit is that the words are *of colours*, and sometimes the colour matches the word (red appearing in red ink) but sometimes they don't match (red appearing in blue ink). The test is considered to measure selective attention and is used in both psychological and neuropsychological circles. Both the ACC and the dorsolateral prefrontal cortex (dlPFC) are activated during this task. The ACC is responsible for selecting the correct response and focusing attention appropriately.

The ACC also has roles in many autonomic functions that are part of the control system that acts mostly below the level of consciousness to control things such as blood pressure and heart rate regulation, in addition to rational cognitive functions such as decision making, empathy, impulse control, emotion and reward anticipation. The dorsal and ventral parts of the ACC are connected to different other areas of the brain and their functions reflect these connections. The dorsal part (dACC) is connected with the prefrontal cortex, parietal cortex and motor system. The ventral part of the ACC (vACC) is connected to the amygdala, nucleus accumbens, hypothalamus and anterior insula. The vACC is responsible for evaluating the salience of emotion and motivational information.

The ACC has a lot of a particular type of neuron called 'spindle cells'. These cells have only one dendrite. We believe them to be important in the intelligence levels of humans as they are linked to the quick processing of information in other mammals. The ACC may be involved in linking between the conscious and unconscious.

A recent study highlighted the involvement of the ACC in mental fatigue (Choo, 2005). Almost everyone experiences mental fatigue or exhaustion at one time or another. At the National University of Singapore, researchers used functional magnetic resonance imaging (fMRI) to look at the brains of men and women aged 19 to 25. These individuals were deprived of sleep for 25 hours and given a simple task repeatedly throughout that time period. It was found that this gradual increase in mental fatigue correlated with decreased activity in the ACC. We know this area to be important as an interface between motivation, cognition and action. It is also linked to the utilization of reinforcement information to control behaviour. So during mental fatigue when we feel lethargic and our thoughts feel slow it is possibly because our ACC is suffering from decreased activation!

Another study looked at the actual way that meditation can reduce anxiety. Mindfulness meditation was the style used here and the participants of the study all had no previous experience. They went through four

20-minute classes to learn some mindfulness meditation. As you'll read in Chapter 28, this involves focusing on breathing and sensations within the body. It also means non-judgementally evaluating any distracting thoughts or emotions. The scanner that looked at the participants' brains before and after they meditated did arterial spin-labelling MRI. The participants were also asked about their anxiety levels before and afterwards. The best results showed a reduction of 39 per cent in anxiety levels. This correlated with increased activity in the ACC and also the ventromedial prefrontal cortex (this area is involved in controlling worrying and executive-level function).

Why is it important to me as a coach?

ACC is often referred to as the conflict monitor or detector. As you've already seen, the ACC connects to different parts of the rest of the brain and can be considered, in simple terms, to connect the thinking and emotional brain.

Being able to talk to a client or colleague in terms of the effect that mental fatigue is having on their brains can sometimes add a degree of distance to problems, which can make them easier to handle. As a manager, if you notice that a team member is struggling with mental fatigue regularly, and it might affect their health and well-being, then a conversation could be useful. How this is approached is obviously of great importance. Simply starting by sharing the above study could be a good way to illustrate that it isn't anything a person is doing wrong or should be pushing through. Rather, it is more like taking care of a machine that has certain limitations.

Something else of interest is a variation on the original Stroop task involving a mix of negative emotional words such as 'grief', 'violence' and 'pain' along with neutral words such as 'clock', 'door' and 'shoe'. The words are in various colours and the aim is to name the colour of the word. Many experiments have shown that people who are depressed take longer to name the colour of the negative words than the neutral ones. This highlights the conflict between the emotional relevance to the individual (dependent on their mental state) and the word.

Often we learn a lot in biology about when things don't work normally. Another area of research to be aware of as a coach is obsessive compulsive disorder (OCD). While few clients will actually suffer from this debilitating condition, it tells us more about the role of the ACC. In people with OCD low levels of glutamate activity have been noticed in the ACC whereas, in other areas of the brain, high levels have been present. Lower grey-matter volumes have also been noted in the ACC. In other pathologies, when lesions are present in the ACC people have difficulty performing the Stroop task, exhibit emotional instability, have an inability to detect errors and experience inattention.

What can I do with a client now having understood this?

The ACC may turn out to be the centre of free will in your clients (and you, of course!). Francis Crick (co-discoverer of DNA) proposes the ACC as a front-runner candidate. With this area of the brain there may be less to 'do' on a session-by-session basis and more to have an overall awareness of.

There is a great experiment that involves a screen, a dot and a black box in the centre of the screen. The dot travels from the left to the right of the screen at a constant velocity. When it reaches the black box it disappears behind it. When it emerges you have to state whether it travelled behind the box at the speed you expected or more quickly or more slowly. The ACC is instrumental in detecting this potential unexpected occurrence.

Expectations are powerful things and our ACC is expert in matching whether what we expect is what we get. If we speak to our boss after just handing in a report we've worked really hard on, and are expecting a 'thank you' or some praise for efforts but instead get a brush-off, then we need to make adjustments.

Our ACC is dedicated to drawing our attention to what is different from what we expect. Perhaps you've had the experience of receiving an e-mail from a colleague and something feeling not quite right. Maybe you brush it off and move quickly on to the next thing on your to-do list. It may be that your ACC had picked up on something important. Perhaps you walk into a meeting and something just isn't what you were expecting.

The moral of the ACC if it had one would be to pay attention to error detection. We also know that the ACC plays a general role in behavioural adjustments. Things like credibility strongly influence the judgement of a person about whether something is true or false. Economic decisions can also depend strongly on framing. Some studies provide evidence that the framing-induced judgement biases are linked to emotional information that is integrated into any cognitive decision-making processes.

Another ACC-friendly option is to reassess a situation. How we interpret a situation – as coaches we know this – is in a large percentage down to us. Utilizing your skills as a coach to reappraise or reassess a situation can be very useful.

Hypothalamus

07

Responsible for various metabolic processes and synthesizing and secreting neurohormones.

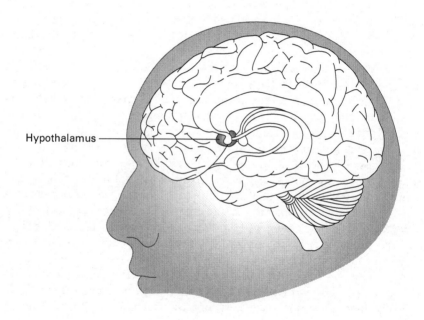

Hypothalamus

What is this brain area?

The hypothalamus is a small area of the brain whose roles include linking the nervous system to the endocrine system via the pituitary gland. The hypothalamus is responsible for various metabolic processes and synthesizing and secreting neurohormones. It is one of the brain areas that it is important to know about for a complete picture, but rarely would direct reference be made to it during a coaching session.

The part of the hypothalamus called the ventromedial nucleus is responsible for the control of food intake. There are several hypotheses that could explain the regulation process. One of these is that adipose tissue (fat, to most of us) produces a proportional signal that acts on the hypothalamus to decrease food intake and increase energy output. Evidence supports leptin (a hormone) acting on the hypothalamus in this way. Another hypothesis is that certain gastrointestinal hormones inhibit food intake. These hormones are released when food gets into the gastrointestinal tract (the stomach being a part of this) and signal to the brain to produce feelings of satiety. A third hypothesis suggests that glucose utilization in the neurons affects our natural food intake. So when we have low blood glucose levels we feel hungry. Another theory suggests that a decrease in body temperature below a set point makes us feel hungry, and above a set point inhibits appetite. Ever felt hungrier in winter and less interested in food in summer?

Recent research has given the media a field day. Newspapers state that the 'secret to living longer' has been uncovered. The research has been done in mice, and the results were impressive. Dongsheng Cai, a researcher at the Albert Einstein College of Medicine, said: 'If we just activated this pathway in the hypothalamus, it accelerated aging, if we inhibit this pathway, we can slow down aging. There were lots of assessments that showed inhibiting it led to an increase of lifespan by roughly 20 per cent. So, that's pretty remarkable' (Dongsheng *et al*, 2013). The researchers were working with a protein complex involved in inflammation in the hypothalamus called NF-kB. The natural process of inflammation in our bodies has a positive intention, but too much of it can be a problem. It was also discovered that GnRH (a brain peptide) is important in neurogenesis. The idea is that by increasing GnRH there would be an increased stimulation of new brain cells and this would work synchronously with reducing NF-kB to reduce inflammation.

Why is it important to me as a coach?

The hypothalamus is important in regulating hunger, aspects of parenting and attachment behaviours, thirst, fatigue and sleep. It is considered part of the limbic system and connected to the amygdala. It responds to olfactory stimuli (including pheromones) and stress. It is also involved in the mediation of emotional responses.

For any coaches working with people trying to lose weight, due attention should be paid to how our brains create and respond to hunger and satiety. Some of the hypothesis will fit with your foundational understanding of hunger; however, in recent times with a variety of diets hitting the market-place sometimes these basic principles seem to get a little lost. If a client is struggling, it may be worth returning to basics and looking at what is happening. It would also be wise to bear in mind what we know about willpower and self-control. Look at Chapter 24 on willpower for some ideas.

What can I do with a client now having understood this?

There are not any obvious things you can do with a client having understood the primary functions of the hypothalamus unless you are perhaps working on weight-loss goals.

Hippocampus

08

Has roles in the consolidation of information from short- to long-term memory.

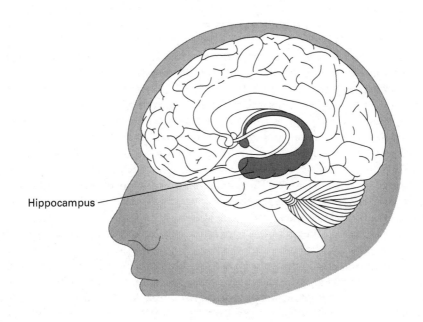

Hippocampus

What is this brain area?

The hippocampus can be found in the medial temporal lobe of the brain. Visually it resembles a seahorse and its name, in Greek, comes from components meaning 'sea monster'. It is considered part of the limbic system and has roles in the consolidation of information from short- to long-term memory.

Long-term potentiation (LTP), one of the mechanisms by which memory is stored in the brain, was first discovered in the hippocampus and has been studied frequently in that area of the brain (a great area to study LTP because there are clear layers of densely packed neurons). LTP involves altering the strength of connections between neurons (we'll learn more about this in Chapter 16 on neuroplasticity). Donald Hebb is credited for the roots of the simplified phrase: 'Cells that fire together, wire together' (Hebb, 2002).

More recently, quite a famous study looked at taxi drivers. This study was carried out at University College London by Eleanor Maguire and colleagues in 2000 (Maguire *et al*, 2006). It was known that people's hippocampi are more active when they are successfully navigating, as this had been tested in simulation tasks. It is also thought to be involved when we find shortcuts or alternate routes. The study at UCL looked at black cab drivers' brains and compared them to normal drivers' brains. What they saw was that the cab drivers had bigger hippocampi. The question then becomes: are people with bigger hippocampi naturally drawn to jobs where they are naturally talented – or does navigating and finding shortcuts daily make the hippocampus grow bigger? This study did show that the amount of grey matter correlated with the length of time the cab drivers had been in this role. (Specifically what was found was that the posterior portion of the hippocampus grew at the expense of the anterior portion, but this doesn't appear to have any detrimental effects.)

Interestingly, rather than a correlation between the overall size of the hippocampus and an individual's ability to recall things, a recent study showed that it was instead the relative contributions of the posterior and anterior parts of the hippocampus. In tests with normal individuals it was found that a larger posterior hippocampus, and a smaller anterior hippocampus, was linked to better recall. It has also been shown that the link between the posterior hippocampus and recollection depended on interactions with other parts of the brain.

There is some evidence that post-traumatic stress disorder (PTSD) and cortisol levels can affect the size of the hippocampus. For example, a study looked at 15 children aged 7–13 who were suffering from PTSD over a time frame of 12–18 months. They measured the children's bedtime cortisol levels (which is a marker of stress) and found that those with more severe symptoms and higher levels of stress also had reductions in their hippocampal volumes. We know that cortisol, from the hormonal group of glucocorticoids, has been shown to kill hippocampal cells in animals.

A study on elderly people over a 10-year period gives us confirmation that the hippocampus decreases in size as a result of depression (Heijer *et al*,

2011). The study concluded after scanning individuals that people with a smaller hippocampus to start with were not at higher risk of developing depression. Instead, it is the depression that causes a decrease in the size of the hippocampus.

A great question is often: 'can we improve this area of the brain?' A linked question when we talk about areas that are involved in memory is often: 'does it have to go downhill as we get older?' Research helps us answer both of these questions. The short answer is that 'yes' we can improve hippocampal function and 'no' your memory doesn't have to go downhill as you age. Studies have reliably shown a relationship between the size of the hippo-campus and memory performance. A study showed that aerobic exercise could increase the size of the hippocampus in adults (this study involved people 55–80 years old) (Erickson *et al*, 2010). Another study looked at physical fitness in children aged 9–10, which showed that those who were fit performed better on memory tests and had larger hippocampal volumes.

As is often the great thing within science, sometimes studies show one thing and then another study shows something to the contrary. It often takes time and many studies to get to a conclusive point. This is just something that we need to be aware of and exercise caution around. So the proviso to the above statements are that some studies have shown a loss of neurons in the hippocampus of elderly people, but further studies, using more precise techniques, showed minimal differences between older and younger people. Again, some MRI studies have shown the hippocampus decreasing in size, while others have not shown this.

Why is it important to me as a coach?

It is generally agreed that the hippocampus has three main functions: inhibi-tion, memory and space. What clients remember and what they forget has potential ramifications on their levels of fulfilment and ability to reach their goals. The hippocampus has high levels of glucocorticoid receptors (cortisol falls into this category of hormones) and so it is susceptible to long-term stress.

In an organizational situation people can often be resistant to change. We often find that people's memories of what has gone before in that organ-ization are very important to them. Being aware of a few things around memories can be really useful. First, just because something happened does not mean everyone will remember it the same way. This seems very obvious, but is often forgotten when people are defending their recollection of some-thing. Second, people seem to develop an attachment to what has gone before and can seem to take it as a personal criticism if there is a suggestion of how things can be improved in the future. It might be as simple as giving people the opportunity to voice their historical experience and perspective on the new suggestions.

What can I do with a client now having understood this?

There are several things that, although they do not have concrete studies addressing the outcomes, will not do any harm, have other benefits and may in the future be conclusively proved:

- regular aerobic exercise;
- stress reduction to keep cortisol levels from getting chronically high;
- give your global positioning system (GPS) a break and engage your brain.

PART TWO
Brain chemicals

Discover the communication molecules of your mind

We have explored the key areas of your brain that you need to be aware of, and now we turn our attention to the chemicals that are at work in this amazing organ. What actually are brain chemicals? We use the term here to encompass a range of chemicals that have an effect on the brain. Some also have effects in other areas of the body. To get really sciency for a moment, let's zone in on neurotransmitters, which are the chemicals that *transmit* a signal from a neuron (a brain cell) to another cell across a synapse (a gap). As we look at more of these basic terms in Part Two and Part Three the picture will get clearer and your understanding will increase. A great way to learn anything new is to layer-in information. During our programmes often participants find it reassuring that knowledge of the whole picture is not expected to be crystal clear at the start.

So we can imagine these neurotransmitters to be packaged up into little balls called synaptic vesicles. When instructed, they are released into the synaptic gap so that they can bind to specific receptors on the other side of the gap. If you did A level or A2 biology you would have probably covered the lock-and-key mechanism that we think of as underpinning receptors binding with neurotransmitters. Neurotransmitters include glutamate, dopamine, serotonin, oxytocin, adrenaline and noradrenaline.

Neurotransmitters such as dopamine, serotonin and oxytocin can also act as neuromodulators. This means that they are secreted by a small group of neurons and move through the nervous system affecting lots of neurons.

Some chemicals we cover in Part Two are not neurotransmitters, they are hormones. Cortisol is one such example. It is included here because it has such important actions within the body and on processes linked to the brain – and for a good overview it needed to be mentioned. In fact, we'll take a look at cortisol first.

Cortisol

Famously known as 'the stress hormone' because it is released in response to stress.

What is this brain chemical?

Cortisol is a steroid hormone that is secreted by the adrenal glands (you'll find these just above your kidneys). Generally known as 'the stress hormone' because it is released in response to stress, it has a wide range of functions in the body, including:

- glucose metabolism;
- blood pressure regulation;
- blood sugar level maintenance (through insulin release);
- immune function.

You'll probably have met cortisol through its formal name, hydrocortisone, which you can buy in a cream to treat rashes and eczema.

Why is it important to me as a coach?

In simplified terms, cortisol is involved in both eustress (good stress) and distress (bad stress). Eustress can be thought of as the motivating type of stress that can help you feel challenged, productive and motivated. Distress feels like there is tension and tends not to feel positive. Short-term elevated cortisol levels can affect efficiency, effectiveness, productivity, interpersonal skills, mental state and much more. This corresponds with eustress. Prolonged elevated cortisol levels can impact quality of life (lowered immunity so potentially getting sick more often, higher blood pressure and increased abdominal fat) and impair cognitive performance. This is the distress we tend to be more aware of. Eustress is linked to a heightened state of arousal that primes you to take action. Once this action has been taken your cortisol levels can return to normal. Distress is not so linked to action and often this results in the cortisol remaining in the body – and us feeling generally anxious.

Cortisol often has a bad name, but it can have some great effects. It is released when the body goes into the 'fight-or-flight' state. When you think you are under threat cortisol enables you to have:

- a burst of energy;
- heightened memory functions;
- lower sensitivity to pain;
- elevated blood pressure.

These are all important to enable you to respond when you need to.

An interesting study performed at Northwestern University looked at the cortisol levels throughout the day in older people (Adam *et al*, 2006). These people were also asked to note down whether they felt lonely, angry or tired. Normally when we wake up in the morning our cortisol levels are low, then about 30 minutes later we get a big boost. Throughout the day it reduces until it is at a low again around midnight (or 3–5 hours after the person goes to sleep). The results showed that those people who went to bed feeling lonely, sad or overwhelmed got a bigger boost of cortisol the next morning than other people in the study who didn't feel these emotions. Emma Adam, one of the researchers, suggested this was an adaptive positive intention of cortisol to give these people the boost they needed to get up and out into the world so that they would increase their chances of connecting with other people.

This study also showed that angry people's cortisol levels remained higher throughout the whole day. From the other perspective, those who didn't get a big boost of cortisol in the morning felt tired throughout the day.

Liz Dunn, a social psychologist, did an interesting study that involved cortisol (Dunn *et al*, 2008). The team gave individuals $10 and told them that they could give as much of it away as they wanted, including nothing (keeping it all for themselves). As they had expected, they found that the more money people gave away the happier they felt. The more of the $10 they kept, the more shame they felt. The team was also able to measure the cortisol levels of the participants and found that the more shame people reported feeling, the higher their cortisol levels rose.

What are some scenarios where knowing about this would enable me to best serve my client?

In the working environment there are many occasions when cortisol levels increase. Any time that an individual's primary response might be to fight or flight it is likely that cortisol could be released. For example, if in a meeting they are being asked to justify their actions or thoughts; or when working with colleagues they find difficult. If they share that they are finding things troubling and their mood is lowered as a result of a certain situation, this could be an indicator. Obviously we don't know for sure whether someone will have elevated cortisol levels, but there are some standard scenarios that make it more likely.

Caffeine is known to increase levels, as is sleep deprivation. So any new parents are likely to be suffering a double dose! Severe calorie restriction has been linked to higher baseline cortisol levels. Travelling by train has also been linked to cortisol levels – the longer, more unpredictable and effortful the train journey, the more cortisol is likely to be produced. So when we ask people 'how was your journey?' we are likely getting some indicators as to their state, both in how they answer the question (we've all seen someone in a state of stress) and in what they answer. In an ideal world, someone who has had a tricky journey would be given a hug to help release some oxytocin – acceptable with friends but not always appropriate with clients or from a manager.

As a coach we might be able to see things that are triggering a stress response, but which our clients cannot see themselves. Similarly, just because a client cannot identify a trigger this does not mean they are not stressed.

What can I do with a client now having understood this?

Clients who understand what cortisol is – what effects it has on their body, when it is released and what they can do to decrease it – are empowered. Taking the time to step into the teacher role as a coach and share some of the basics around cortisol can be useful. Equally can being fully present for your client, being non-judgemental and giving them the attention they deserve.

With links to both personal and professional challenges often clients can spiral deeper when they haven't realized what is happening. It is great to ensure that your clients have access to stress-reduction techniques or training. Mindfulness-based stress reduction has been used in several studies now and has shown great results for the individuals involved in reducing stress levels. A meta-analysis of mindfulness-based stress reduction trials in 2011 called 'Mindfulness-based stress reduction and mindfulness-based cognitive therapy – a systematic review of randomized controlled trials' is a good reference source (Fjorback *et al*, 2011). There are lots of qualified teachers of this practice out there who would make good guides and role models for those you work with if you are not yourself best placed. This practice offers lots of additional benefits too, which we focus on in Chapter 28.

Talk to your client about their physical activity levels. Regular physical activity – either in the form of an intensive session such as swimming, an aerobics class or running, or in smaller chunks such as jogging up the stairs or taking a power walk at lunchtime – can all lower the overall cortisol levels.

Explore the social connectivity of the person you are working with. How are they connecting with others and how deep are these connections? Do they have between three and five close friends who they can call on and share important things with? Are they getting daily some connection with people?

Consider asking your client to keep a laughter diary to record how often they laugh during a normal day. Dr William Fry has found links to laughter and lowered levels of stress hormones. Some people might benefit from watching a funny clip each day, reading a joke or, even better, meeting with friends and sharing things that make one another laugh. In a corporate environment consider the culture. Do people laugh? Do people have a sense of humour? Could small steps be taken to help people relax enough to find things funny?

The place of music in life is another consideration. Listening to music has been shown to lower cortisol levels. So it could act both as an immediate support if a person is aware they are feeling in need of calming (this doesn't necessarily mean listening to calming music, anything enjoyable is useful) or listening a couple of times a day as maintenance.

Dopamine

Involved in how we think, behave, our feelings of motivation, reward and attention.

What is this brain chemical?

Dopamine is a neurotransmitter. It is a chemical that transmits signals between brain cells (neurons).

Why is it important to me as a coach?

Dopamine is involved in:

- how we behave;
- how we think;
- our ability to move;
- our feelings of motivation;
- reward;
- punishment;
- working memory;
- attention;
- learning.

Dopamine isn't made by lots of neurons, but the area of the brain called the substantia nigra is home to some dopamine-producing neurons. Another area is the ventral tegmental area (VTA). The dopamine does travel all over the brain. When something good and unexpected happens, such as being given a present, an ice cream, or praise in front of others, then dopamine neurons in a part (dorsal) of the VTA are activated. At this point you want to keep doing whatever triggered the release of dopamine. You feel motivated. When something bad happens, dopamine neurons in another part (ventral) of the VTA become very active. When you expect a reward and it doesn't occur then dopamine neurons are depressed. Dopamine is well known for its role in learning about rewards: learning from bad things stopping is similar to learning from a normal good reward. Dopamine is also important in the frontal lobes for controlling the flow of information from other areas of the brain.

University College London (UCL) is responsible for a couple of interesting experiments investigating decision making. The first involved using imaging techniques to detect a signal in the brain that was linked to how much someone enjoyed an experience. It was then found that the signal could predict the choices a person made. The researchers suspected that this signal was dopamine, so they set up an experiment to see what happened if they tampered with the dopamine system (Sharot et al, 2012). A group of participants were given a list of 80 holiday destinations from all over the world to rate from one to six. They were given a sugar pill, to provide a control portion of the experiment, and asked to imagine themselves in half of the destinations (so individually in 40 of the original destinations). The participants were then given L-Dopa, which has the effect of increasing dopamine in the brain and is used typically to help sufferers of Parkinson's disease as it helps reduce tremors, rigidity, slowness and improve balance. These healthy individuals were then asked to imagine themselves in the other half of the destinations. They rated all the destinations again, then the next day had to choose where they would go out of paired lists of holidays. The increased dopamine made people choose the holidays they had imagined when more of it was flowing through their brains, so they chose more from the second list. The leader of the experiment, Dr Tali Sharot, said: 'Our results indicate that when we consider alternative options when making real-life decisions, dopamine has a role in signalling the expected pleasure from those possible future events. We then use that signal to make our choices.' So, higher levels of dopamine makes us more likely to rate something favourably and subsequently choose it.

Another experiment carried out at UCL, this time led by Professor Ray Dolan, showed that increased dopamine makes us more likely to go for something with instant gratification rather than a more beneficial, but slower to be delivered, reward (Dolan et al, 2010) – this obviously has key implications for coaches. Participants were tested after being given a placebo, and also after being given L-Dopa. The test involved making several choices between 'smaller, sooner' or 'larger, later'. For example, they could choose to receive £15 in two weeks' time or £57 in six months' time. Dr Alex Pine, who was involved in the study, said: 'Every day we are faced with decisions that offer either instant gratification or longer-term, but more significant reward. Do you buy your new iPhone today or wait six months till the price comes down? Do you diet or eat that delicious-looking cake? Do you get out your books to study for a future exam or watch some more TV?' (Dolan et al, 2010). The results of Dolan et al's experiment showed that each of the participants chose more of the 'smaller, sooner' options when they had more dopamine in their brains. This was great for the experimenters as they were only paying out the sum of £15 in two weeks rather than £57 in six months!

What are some scenarios where knowing about this would enable me to best serve my client?

Imagine the common scenario where someone you are working with does not have the best diet. They frequently eat and drink things that are not great for them. Perhaps they enjoy them and feel that they work hard and so deserve these 'small vices' in their life. As usual, depending on your relationship with them you may or may not choose to broach something that someone has not directly brought to you as a challenge. If productivity, effectiveness or efficiency are being addressed in any way, however, and your client asks for your input, diet could be considered fair play.

Dopamine receptors need more stimulation in order to trigger dopamine when we become addicted to something. In corporate environments it is often the norm for people to drink copious amounts of coffee. Caffeine, along with saturated fats and refined foods, are not great for brain function. When we eat a lot of junk food our dopamine production is inhibited. This can have a direct impact on our productivity. As time goes by we can need more caffeine to get that same 'hit' – and this negative spiral is not ideal. This is a scenario for managers in the role of coaches to be aware of, but which also applies to coaches working in the personal arena.

The environment that people are in and the behaviours they are directly or indirectly encouraged to participate in are very important. Two classic pieces of advice have great scientific underpinnings to them. Often by sharing not just 'what' people should do but also 'why' they should do it yields better results. Advice such as 'do aerobic exercise' is commonplace, good old-fashioned reliable advice. Often, though, it is not given the credit, and does not result in the action it should. For some people it helps to know why exercise could be good for their productivity. Exercise increases our blood calcium levels, which stimulates dopamine production and uptake.

What can I do with a client now having understood this?

Teaching your clients how to trigger a dopamine boost from just thinking about the small tasks that are in alignment with their goals will make it much easier for them to follow through with their decisions. For example, when we think about cake many of us experience a dopamine release, which will motivate us to seek out cake! Eating the cake will also trigger a release, programming us to want it again. We've just learnt that having cake makes us feel good. For most people, this isn't ideal on a regular basis. As with any long-term goal you need to be mindful of your short-term actions.

When we think about clearing some e-mails many of us experience that same dopamine release, this helps us want to just sort through a few, perhaps replying with a shorter reply than it deserves just so we can delete or file it. Once we reply to one we often want to send another and another in order to feel that mini buzz when we click 'delete' – so that it is no longer in our inbox and lobbying for our attention. We often find ourselves 'doing e-mails' when there are other things that we could more beneficially give our attention to.

Simply being aware of this and how easily influenced we are by our neurochemicals is useful to your clients. Getting them to practise bouncing from an unhelpful thought, which triggers dopamine, to a useful one is a great skill to strengthen. So perhaps thinking about doing some sit-ups, for example, and linking that in their mind to the bodily figure they are proud of and enjoy could soon trigger that release for them. In essence we need the 'smaller, sooner' rewards to come from things that are in alignment with the 'larger, later' goals. This will likely take an investment of time and energy.

Oxytocin

11

Involved in social behaviour, increasing trust, decreasing fear, increasing generosity and also cognitive functions.

What is this brain chemical?

Oxytocin is a neuromodulator that is released from the pituitary gland. It is most well-known for its role in trust, thanks to the work of Paul Zak, and also is involved in behaviour and cognitive functions.

Oxytocin has the effect of:

- reducing blood pressure;
- reducing cortisol levels;
- increasing pain thresholds;
- an anti-anxiety effect;
- stimulating positive social interactions.

It used to be thought that oxytocin was a hormone with important effects only for women around childbirth and breastfeeding. Now the additional important and wide-ranging effects in social connectedness are more recognized.

Why is it important to me as a coach?

Oxytocin has the effect of suppressing the activity of the amygdala. You'll remember that this brain region detects threats and processes fear. One study used fMRI imaging to look at this area of the brain after giving men either oxytocin or a placebo before they performed a task that involved sorting pictures of angry or fearful faces and threatening scenes. The scans showed, as predicted, lower activity in the amygdalae of the men who had inhaled oxytocin.

Oxytocin has been implemented in promoting ethnocentric behaviour, which is so important in whether people are trusted and empathetic towards people considered 'in' or viewed with suspicion, and the rejection of those considered 'out' of groups. While judging others relative to your culture is considered natural it can also have negative connotations. In organizations it is important to be aware that it can occur. Teams that look down on people from other teams is not uncommon.

A study that looked at oxytocin's 'rival', testosterone, showed some interesting results. It involved men having their normal baseline level of testosterone measured and then them playing the 'ultimatum' game after being given a boost of testosterone through a product called AndroGel, a hand gel. Six weeks later they would play the ultimatum game again but with an ordinary hand gel (they wouldn't know which they were getting or when).

You may be familiar with the ultimatum game, which involves two people and some cash. In essence, player one gets to propose how to divide a sum of money between the two players. Player two gets to accept or reject this proposal. If it is rejected then neither player gets any money; however, if accepted, then both players get to keep the money according to the split proposed. The game is only ever played once so that reciprocity doesn't affect it.

The result in this case was that when men had their testosterone levels artificially raised they were 27 per cent less generous towards strangers than when they had their normal levels. When playing the ultimatum game it was shown that oxytocin increased people's generosity by around 80 per cent. The mechanism at play here is thought to be an increase in empathy during perspective taking.

Another interesting study (this one has less direct applications... but it's nice to have a broad understanding of the fantastic effects of oxytocin) showed that oxytocin may play a part in supporting fidelity within monogamous relationships. When men who were in monogamous relationships were put in a situation with an attractive female they chose to stand at a certain distance from her (unconsciously and unaware this was being measured). The men who had been given oxytocin via a nasal spray stood 10–15 centimetres further away from her than the monogamous men who hadn't received the nasal spray. So before your partner leaves for work a hug to release a bit of extra oxytocin might be a good plan!

Finally it is believed that oxytocin helps the processes of learning and memory specifically for social information.

What are some scenarios where knowing about this would enable me to best serve my client?

If you are an empathetic coach then you will be releasing oxytocin, you'll also probably release dopamine and feel rewarded from connecting with your client, making it more likely that you'll feel empathy again. Whether you coach over the phone or in person you are likely to be building a bond with your client and the effects of oxytocin help this.

Some coaching relationships can need a little help on the bonding front. For example, in a work setting where a manager has been told they need to 'coach' one of the team members there can sometimes be barriers to this relationship being set up optimally. Perhaps the person does not want to be coached. Perhaps they do not want to be coached by their manager. Perhaps they feel like they have been singled out and they may have a problem with that. There are lots of things that could make it much more difficult for a productive relationship to be set up between coach and client.

This could also be the case when an external coach is brought into an organization. Sometimes people do not understand how coaching works, thinking perhaps it is for remedial purposes only. There may be a sense that someone from outside cannot understand internal matters. In short, there can be a long list of reasons why people may not be open to coaching. This gives you, the coach, a problem. Obviously some issues need to be verbally addressed in the proper setting and using appropriate channels. Another approach to overcoming this problem is to focus on raising the oxytocin levels of the individuals you are called to work with.

There are many ways that have been researched that raise people's oxytocin levels. There are also many more things that may do so that have not yet been fully researched. Here we will focus on activities that have the backing of solid research. Consider how you could share a meal with someone you are working with. Even better, if possible, is to cook the meal for the person. The act of eating can be calming and help us to bond with others. Good wine and good conversation can be really conducive to releasing oxytocin and building good foundations to working relationships.

Perhaps think of an appropriate gift you could give someone you are working with. Would a stimulating book (on a topic they are working on), a thoughtful experience (something that links to a hobby), or a memory of something they are proud of (a photo of an achievement) be something you could make happen? Receiving gifts has been shown to raise oxytocin levels.

When you are face-to-face with a client, and indeed anyone, you have the opportunity to give people your full attention. This opportunity is rarely taken as seriously as it should be. The power of really looking at someone and seeing and hearing what they are communicating to you is very valuable. Taking the opportunity to really observe and tune in to your client could be very worthwhile.

As a coach there may be scenarios where you want to increase the amount of oxytocin you release. Some of these activities also, through various means, trigger the release of oxytocin in those you are with too. You could start by hugging someone, which should also trigger the release of oxytocin in them too. Some professionals are now choosing to let people know that they tend to hug rather than shake hands.

If you are a regular meditator, and even if you are not, then you might consider the form of meditation called 'metta'. This form of meditation involves you sending or radiating loving kindness to others and has been shown to foster social connections better than mindfulness meditations (in as short as seven minutes). In this meditation you progressively send this loving kindness to yourself, a good friend, a neutral person, a difficult person, all four of the previous people equally and then gradually the entire universe!

In any form of personal relationship coaching, so we're not talking about work relationships here(!), another fairly reliable way to trigger the release of oxytocin is to have an orgasm – it is important to note, however, that this has to be with another person you care about in order to trigger the oxytocin release. This can be really useful for helping couples to reconnect and build on what is good. Another way of doing this is through massage. This form of touch is great at releasing oxytocin both in masseur and recipient.

Let's now consider the insights that we received from the study looking at the effects of testosterone and oxytocin. Increased testosterone levels made men less generous. When would the person you are working with be generous, or not be generous? It is possible that a sales coach, for example, might want to share this study with a client. Perhaps they have heard about practices that are described as getting individuals 'in the zone' and in fact are increasing testosterone levels. Other activities such as watching or listening to a favourite sports team win a game can release testosterone, lifting weights can lead to a peak testosterone level 48 hours later, and getting sunlight on your skin can also increase testosterone levels.

Oxytocin has also been shown to improve wound healing. While this scenario may not come up often, it could be potentially very useful to know. A study was done using heterosexual couples; one was given a small blister wound. The couples then participated in a structured support interaction task. They measured the healing of the wounds and also the oxytocin levels (among other compounds) over the coming days. They found that higher oxytocin levels were associated with more positive communication behaviours during the structured interaction task. Also, among these couples, the person with the wound healed faster than the individuals in a lower oxytocin quartile.

What can I do with a client now having understood this?

A starting point here may be to help your client to understand when they release oxytocin, why and what effect it has. This could benefit them in a range of situations, especially to strengthen social connectedness.

For some people, looking at their social media usage may be of value. People may reach out through social media in an attempt to boost their oxytocin levels. Some have said that this is not an effective way to feel more connected to others. However, the research shows that people *do* experience an increase in their oxytocin levels when they use social media. Coaching enables people to explore whether their intentions are being met without any negative side effects. For example, is an entrepreneur who works alone getting distracted and suffering from lower productivity levels because they are trying to reach out and feel connected to others through their use of Facebook? If so, perhaps more of a balance needs to be struck.

You might want to suggest that your client does something a bit daring with someone they want to build more of a connection with. Perhaps it is time for a work trip to a theme park? Riding on a thrilling rollercoaster with the boss next to you can be a bonding experience for both. Maybe your client wants to do a sky dive as a reward for achieving a goal. Make the most of the opportunity and suggest they invite along with them someone they want to be even closer to – and then when their oxytocin levels are raised from doing the jump they can enjoy the increased bonding with that person when safely back on the ground.

Adrenaline

12

Most well known for its role in the fight-or-flight response.

What is this brain chemical?

Adrenaline is also known as epinephrine and in the future it is likely everyone will favour the term epinephrine. It is both a hormone and a neurotransmitter. You have probably heard about adrenaline and its role in the fight-or-flight response. It is important for regulating our heart rate, blood vessel and air passage diameters and has metabolic impacts.

Adrenaline acts on nearly every tissue in the body. In the heart it increases the rate at which it beats. It can cause both vasoconstriction and vasodilation (constricting and dilating the muscular walls of vessels). In the lungs it increases the rate at which we breathe. In the liver it stimulates glycogenolysis, which means the breakdown of stored glycogen to glucose and glycogen. It triggers lipolysis, meaning the breakdown of lipids into glycerol and fatty acids, giving us fuel to use. It also can cause muscle contraction.

The classic way that people remember the basic effects of adrenaline are to consider what we want to be able to do if a lion appears in front of us. By our heart pumping faster there is more blood with all its available nutrients being sent quicker around our body. Combine this with us taking in more oxygen by breathing at an increased rate and this means that we are primed for action. Add in the action of breaking down glycogen to make more glucose available and we're ready to run.

Why is it important to me as a coach?

Adrenaline is part of our emotional response to things. This process is controlled by the sympathetic nervous system. Fear is the best studied of the emotions in relation to adrenaline. Studies have shown that when we have higher levels of adrenaline we show more negative facial expressions in response to a scary film, for example, than people with lower levels of adrenaline. The studies also reported a greater intensity of negative memories. The same was not the case for other emotions. So people given an adrenaline boost did not find funny films funnier, for example, or angry films angrier.

Adrenaline can produce retrograde enhancement of long-term memory. When an emotionally stressful event occurs, adrenaline is released. This can modulate memory consolidation. If we think about this, it is very useful in theory for us to better remember the things that we are marking as more important than those that are less important. In practice there can be some complications with this, however. Some of the complications include post-traumatic stress disorder, where adrenaline is thought to play a role in increasing arousal and fear memory.

An interesting effect of adrenaline is 'broken heart syndrome' or Takotsubo cardiomyopathy. This syndrome affects people who have suffered severe emotional or physical stress. Often it is seen in older women after their husbands pass away. Initially the symptoms resemble a heart attack but a full recovery is normally made within days or weeks. The process that researchers believe is occurring is that the body changes its response to adrenaline by switching from stimulating the heart to reducing its pumping power. This is important for medics to be aware of so they do not give more adrenaline. It is also interesting to us that the heart may change its response to adrenaline, hypothetically to protect it from being overstimulated from high levels of adrenaline that are normally released during stress.

Until recently we have associated chronic stress with damage to DNA. In the last few years a mechanism has been proposed that helps to explain how this actually happens. When people suffer from chronic stress they have elevated adrenaline levels. Researchers found that chronic stress lowered p53 levels. This is a tumour suppressor protein and its role extends to preventing abnormalities of the genome. In mice this effect was noticed in just four weeks.

On a more positive note there are lots of great times when adrenaline is also involved. When we are falling in love our body releases a mix of chemicals, including adrenaline, noradrenaline and dopamine. The dopamine is the one that makes us feel euphoric. Adrenaline and noradrenaline make our hearts beat faster and make us feel a bit restless and preoccupied.

Knowing the links between chronic stress, high levels of adrenaline and long-term potential health problems is important. It establishes the stakes that people you work with may be gambling with.

What are some scenarios where knowing about this would enable me to best serve my client?

Imagine you are working with a senior leader in an organization. You have been working with them for some months now. For as long as you have known them they have been suffering from chronic stress. They have raised this on several occasions and each time said that they would deal with it later. You are starting to notice a pattern. By understanding the high stakes that they may be playing with, you are in a position to share this with them. You might invite them to consider the research and whether they are still happy playing on this high-stakes table.

What can I do with a client now having understood this?

In a work environment – so for corporate coaches and managers in the role of coaches – doing an adrenaline audit could be useful. Looking at an average week, what are the times that stand out where an adrenaline surge may be occurring? For some people this will be obvious and they will be very connected to the feelings they experienced linked to an adrenaline surge. At this point the emphasis is on awareness for personal management. So by becoming aware of when, and how often, a big adrenaline rush occurs, people can then consider whether they want to do anything about that.

In some lines of work a regular adrenaline rush is considered normal and individuals are not experiencing any adverse effects from it. Perhaps for people in sales, the thrill of the chase and being on tenterhooks before finding out whether a proposal has been accepted or not is all part of a healthy process. For other people, tense relationships, poor management and unhelpful systems can all lead to adrenaline bursts that negatively affect health and may warrant investigating.

For many clients, adrenaline is only being released appropriately and is not linked to any chronic stress problems. As coaches we have to keep our curious hats on and avoid presuming. Trusting our clients to work things through for themselves can sometimes be a cause of adrenaline release in us!

Serotonin

13

Important for mood regulation, appetite, sleep, memory and learning.

What is this brain chemical?

Serotonin is also known as 5-hydroxytryptamine (5-HT), which becomes more interesting when you know that serotonin is made from tryptophan because we can see the link between one of the precursors and the end product. It is one of the most well-known neurotransmitters and has its reputation firmly established for being important to feeling happy. In addition to being involved in mood regulation it is also important to appetite, sleep, memory and learning. When levels of serotonin are low we can feel low and even have symptoms of depression or anxiety.

Tryptophan, the essential amino acid needed to make serotonin, is found in our diet. Bananas, dates, yoghurt, milk, chocolate, sesame, sunflower and pumpkin seeds, and poultry all have high levels of tryptophan. Some foods are extra special because they have a high ratio of tryptophan to phenylalanine and leucine (two other amino acids) and this ratio is known to increase serotonin levels. So eating dates, papayas and bananas seems like a good plan.

On the other hand, drinking alcohol has the effect of decreasing tryptophan levels. Drinking an average amount of alcohol leads to a decrease of around 25 per cent of tryptophan, which translates to a similar decrease in serotonin. It is suggested that some of the more risky sexual and impulsive behaviours linked to drinking are at least partly a result of the decreased serotonin levels that normally have a role in regulating these behaviours.

While serotonin is released by neurons in the pineal gland of the brain it has effects all over the body. It is involved in regulating the sleep cycle and circadian rhythms (along with melatonin) and is affected by sunlight. Sunlight has ultraviolet light as a constituent part and our skin absorbs this to produce Vitamin D, which is important in promoting the production of serotonin. The exposure to sunlight is something many people find more challenging in the autumn and winter months. The rise of seasonal affective disorder (or SAD) is one of the results we see. Either getting more sunlight or using a well-researched, credible light box can help boost serotonin levels and will hopefully improve mood.

When we say that people have low serotonin activity or levels, what is actually happening inside the brain can be one of several things. One explanation is that your brain is making less serotonin. Another is that there are fewer serotonin receptors or that the receptors there are not receiving the serotonin well. Alternatively it might be that although you are making lots of serotonin it is being reabsorbed too quickly back into the neuron from the synapse, or that it is being broken down too soon.

Why is it important to me as a coach?

Some of the figures state that 1 in 10 Americans are taking an antidepressant. In some cities in the United Kingdom, 1 adult in 6 is on antidepressants. One of the most popular forms of antidepressants is what is known as a selective serotonin reuptake inhibitor, or SSRI. These work by increasing the amount of serotonin in the synaptic gap by inhibiting the reuptake of serotonin back into the presynaptic cell. This means that there is more available to bind to the postsynaptic receptors. Recent studies have looked at the effectiveness of SSRIs in people who are mildly, moderately or severely depressed. People suffering from mild and moderate depression, it was found, received small or no additional benefit when compared to taking a placebo. The results are less conclusive in people with severe depression.

Serotonin and dopamine work together to help regulate the appetite. When we smell food, dopamine is released to increase our appetite. While we are then eating food, serotonin activates a type of receptor that affects dopamine-producing cells. The effect is to stop the release of dopamine, thus reducing appetite. In the absence of serotonin, one is less able to identify when they are no longer hungry and this is linked to weight gain.

Serotonin levels can be raised. Getting enough sleep is important. Coaches should be wary of being too prescriptive on this point, though. What constitutes 'enough' sleep for one person can be very different to what is 'enough' for another person. A change is normally a good indicator of something worth addressing. If a person had been sleeping well for eight hours and recently has been waking up after five, unable to get back to sleep, then it is possible that this could be affecting them.

Some research has been done into massage and serotonin levels. A study was carried out with depressed pregnant woman (Field *et al*, 2004). These subjects received two massages per week from their partners for four months. A 30 per cent increase in serotonin levels was measured. Another study looked at the effect of massage on babies. These babies had depressed mothers, so this was another group where this research made a lot of sense. These 1–3 month olds were massaged for 15 minutes for 6 weeks and their

serotonin levels increased by 34 per cent. There are c
explore a non-loved-one doing the massaging. Overa
studies we could generalize to say that massage appea
levels and increase serotonin and dopamine levels. Th
on people with a variety of medical conditions and
things.

What are some scenarios where knowing about this would enable me to best serve my client?

Being a coach you know that the mind is very powerful. Even thinking about things has a profound impact on our biochemistry. Although as a coach you are probably acutely aware of this, it is worth remembering that not everyone is. Calling to mind something positive – such as a happy memory, an achievement, a great experience – increases serotonin production in the anterior cingulate cortex (ACC). This is an area of the brain just behind the prefrontal cortex that controls attention. On the other hand, if clients find themselves ruminating on sad events then they can expect a decrease in serotonin production in the ACC. Being consciously aware of what is filling your mind and the effect it has on you is important and can dramatically affect how your day goes. Systemizing your thoughts may sound a little drastic but can be very useful in creating useful habits and getting your biochemistry to a place where you can then be on more of a roll. For example, when coaches suggest starting the day with an 'attitude of gratitude'-type exercise this could set people up for a more productive day.

Exercise has been shown to be important both in serotonin production and release. So for anyone who wants more motivation to start exercising, or to continue exercising, this knowledge might be helpful. Typically we focus here on aerobic exercises, which are great for boosting serotonin. However, if your client is a yoga fan then that could work too. There appears to be a difference between choosing to exercise willingly, and exercise that you feel forced into doing. A point for another time, though, is that often to create a new helpful habit you may have to do things that sometimes you're not feeling the love for, but it will come later.

If a client is struggling to exercise willpower to delay gratification in some way then their serotonin activity may be low. It may also be proving difficult for them to create and deliver on well-formed plans. When we consider what is normally necessary to achieve a goal then we can see that low serotonin activity may make this more difficult. Sometimes there are steps that need to be taken before a goal is worked on. Building up an individual to a place where they stand a good chance of achieving a goal can be a really good exercise.

What can I do with a client now having understood this?

Overall, anyone looking to work on a goal should do well to reflect on times when they achieved their goals. Not only is this in itself likely to release serotonin, there may also be things they could learn to help them get into optimal goal-achieving states!

In a corporate environment, however, it could be useful to look at the ways in which it is easy for people to raise their serotonin levels. It is important here to remember that sunlight, massage, exercise and memories of happy events are all great ways to boost serotonin levels. Depending on how and who you coach you might want to explore how they as an individual can take personal responsibility for how they get these activities into their life more. If someone is looking to increase the serotonin levels of others who they work with then they could find creative ways to encourage these activities. Perhaps a team exercise competition, where over the course of a month participants post publicly how often they are exercising. The people who stick to their goals win a massage! Could any meetings be held outside rather than indoors in order to boost sunlight exposure? How do happy, successful events get remembered corporately? As coaches these are some of the questions and thoughts you might be having.

When working with individuals in a personal capacity you have even more scope to hold them accountable to their own levels of serotonin and happiness. As mentioned above, sometimes taking the time to work on preparing someone to undertake a goal is the best way to start. A good plan is to coach a client using lots of open questions to help them identify how they think they can get their serotonin flowing, and cross-referencing this with what the science tells us. Some people know themselves really well, and so trusting what they intuitively feel would be good strengthening exercises for them could also work well.

GABA and glutamate

14

The king and queen of neurotransmitters.

What are these brain chemicals?

Glutamate and GABA are very important neurotransmitters. They are the most fundamental of all the neurotransmitters. For coaches they are a little less easy to position as sexy and relevant but they are included here because they are so key to brain function. The biochemistry that underpins their activity is understandably quite full and there is not space, nor probably the interest from most, unfortunately, to warrant its inclusion. So here is a simple overview of what they are.

Glutamic acid is a non-essential amino acid, meaning that it can be made in the body from essential acids. Officially it is the carboxylate anions and salts of glutamic acid that are known as glutamates. Glutamate has very important roles as a neurotransmitter vital for long-term potentiation (a form of neuroplasticity) and involved in learning and memory. Glutamate is the most abundant excitatory neurotransmitter in humans.

Interestingly, most of us are familiar with the manufactured salt flavour of glutamate in the monosodium glutamate used in a lot of Chinese take-away foods. Glutamic acid is responsible for the flavour we call umami, one of our five basic tastes.

GABA is gamma-aminobutyric acid, the chief inhibitory neurotransmitter (just as glutamate is the chief excitatory neurotransmitter). It acts to inhibit signal transmissions between nerve cells. Glutamine availability impacts the manufacture of GABA. Adequate levels of GABA impacts peace of mind, balance, concentration and restorative sleep.

Why are they important to me as a coach?

Glutamate's involvement in learning and attention is relevant to coaches. Several studies have shown that chronic stress can trigger a decrease in glutamine levels and a suppression of its function in the prefrontal cortex (PFC). This can have a detrimental effect on PFC-dependent cognitive processes, of which we know there are many. This means that there is a very real scientific underpinning to chronic stress, making it more difficult to think, process information, make decisions and all the other cognitive processes that many of our clients need to perform every day.

During acute stress episodes, cortisol is secreted from the adrenal glands. Certain areas of the brain (hippocampus and PFC) have a lot of cortisol receptors. When they are activated an increase in the amount of glutamate is noted. More glutamate in the short term leads to heightened performance in these brain areas, as part of the fight-or-flight response. The glutamate is normally cleared from the synapse but if there isn't enough time between stress phases, as is the case with chronic stress, then the high levels of glutamate can lead to excitotoxicity; damage to cells causing a reduction in brain areas.

A few other interesting facts includes that the more glutamate receptors someone has the higher their intelligence. 'Overexcitability' from glutamate can be controlled by diet reductions in gluten, casein, glutamate (eg monosodium glutamate, or MSG) and aspartate. Alcohol abuse is linked to higher levels of GABA, and higher levels of GABA have been linked to depression.

PART THREE
Foundational
brain concepts

In Part Three we focus on the foundational things that a coach needs to understand about the brain and what this means to them. We look at the core units of the brain, the neurons and synapses. Then we explore the fascinating area of neuroplasticity. When we teach neuroplasticity in depth in the Neuroscience for Coaching programme, coaches often tell us that they find it really exciting. They recognize that the area of neuroplasticity is fundamental to coaching and it offers a scientific explanation for a lot of what they experience in their coaching work.

There are several conceptually useful models that didn't make it into this section because they are scientifically disputed and, while it is possible that others may need revising in future years, we have a good balanced snapshot between usefulness and science today.

The threat response is of particular importance to coaches in a work setting. The problems that arise from the threat response occurring in individuals, and this not being recognized, frequently lead to mismanagement. Lower efficiency, effectiveness and productivity are all side effects of the threat response. Being aware of it gives people the opportunity to act differently.

We also investigate the various types of brain scans that tend to be mentioned in studies that you may be interested in and how they work. This overview is very useful to any coaches wanting a deeper understanding of what is being discovered in the field. Next we reveal what some of the scientific community are saying about the triune brain and what you might like to do as a result of this.

The working memory is the fifth topic in this section and is a good 'general awareness' area to have up your sleeve as a coach. With some concepts it is not likely that you would discuss them very often with your clients, but being aware of them may link into your work in other ways. The hypothalamic-pituitary-adrenal (HPA) axis, by contrast, is something that is being activated in those you work with on a regular basis and so an understanding of it may be of use frequently. Similarly an understanding of mirror neurons may come up time and time again. Vilayanur Ramachandran said in a playful way that 'mirror neurons will do for psychology what DNA did for biology'.

The underpinnings covered in this section provide a firm foundation for the following parts of the book, in which we look at diverse constructs such as trust, willpower and powerful influencers on behaviour.

Neurons and synapses

15

What are they?

The body is made up of cells. Neurons are otherwise known as 'brain cells'. They are electrically excitable cells that process and transmit information. They communicate either electrically or, more commonly, chemically. Neurons have various components to them including a core cell body with a nucleus (like the cell's mini brain), an axon (a long single extension coming out of the cell body), and dendrites (thin structures that also extend from the cell body, but of which there are many and branch multiple times giving the illusion of a 'dendritic tree'). The axon is insulated by a myelinated sheath, which enables electric currents to pass efficiently down the axon. At the end of an axon there is a little bulb, the axon terminal (sometimes called the synaptic knob).

The place where one neuron communicates or transmits a chemical signal to another is called the synapse. They are the specialized connections through which cells communicate chemically. Although we call them a connection, in fact the synapse is also called the synaptic gap. There is actually a gap between one synaptic knob and another dendrite. We can think of this first part, the presynaptic neuron, as having something to communicate and it is the postsynaptic cell that is prepared to listen.

An action potential (a nerve impulse) travels down the axon. When it reaches the axon terminal synaptic vesicles containing neurotransmitters are released. These neurotransmitters then travel across the synaptic gap and attach to receptor sites on the postsynaptic cell. This may or may not generate another action potential here. There is no other option, so it isn't a question of strength of action potential – it is either a stop or go situation, a yes or no, an action potential or no action potential situation. Each neuron can be connected to as many as 15,000 other neurons through these synapses.

Receptor sites that are positioned on the postsynaptic cell are ready for neurotransmitter molecules to bind with them in a lock-and-key-type fashion. When this happens they are telling the cell to do something. Each receptor is specific and will only bind with a specific neurotransmitter. The action is either excitatory or inhibitory, so it either results in the generation of an action potential or it does not.

The process that occurs at the synapse does not offer direct applications for a coach, however a general understanding of the process is a good foundation and is as follows:

1 The neurotransmitter is made, or synthesized. This can happen in either the cell body, the axon or the axon terminal.

2 The neurotransmitter is stored in synaptic vesicles in the axon terminal, waiting.

3 Calcium ions enter the axon terminal during an action potential (a nerve impulse) and this signals to the synaptic vesicles to release their neurotransmitters into the synaptic gap.

4 Once the neurotransmitter is in the synaptic gap it binds to and activates the receptor sites in the postsynaptic membrane (the edge of the postsynaptic cell).

5 Finally, the neurotransmitter is deactivated. Either this happens by it being destroyed by enzymes or it gets taken back into the presynaptic neuron to be reused or destroyed.

For videos that bring this process to life check out:
www.neuroscienceforcoaches.com

Why are they important to me as a coach?

This fundamental process is very important. The formation of memories at its most basic level occurs at the synapse. As neurotransmitters communicate from the presynaptic neuron, across the synaptic gap to the postsynaptic neuron, the connection between the two neurons is strengthened. The analogy we use during the Neuroscience for Coaches training programme is that it is like two people passing a ball between themselves. Each person is a neuron and the ball is the neurotransmitter. Normally the passing of the ball is a little slow and clunky, sometimes it even gets dropped on the floor. However, the more they throw it back and forth, the more the neurons communicate with one another, the quicker they are able to. We talk about the connection between the neurons as being strengthened and the strength results in the storage of information, which results in memory. This process is called long-term potentiation.

Long-term potentiation is described as a long-lasting enhancement in the transmission of signals between two neurons as a result of stimulating them synchronously. It is so important for coaches because it underlies synaptic plasticity, or neuroplasticity as it is more commonly known. This ability for synapses to change their strength is vital for learning and memory.

Neuroplasticity

16

What is it?

Neuroplasticity is the property of the brain to change. That in itself is quite a general property, but nonetheless an important one. There are several different ways that the brain naturally changes. When the strength of the connections between synapses changes we call this synaptic plasticity. This focuses our attention on the location of the change. A little more specifically, long-term potentiation could be occurring (we mention this in Chapter 8 on the hippocampus and Chapter 14 on GABA). Alternatively there could be more or fewer receptors present for specific neurotransmitters, or a number of other things could be different. Action at the synapses is very important within the brain.

Other changes that come under the broad banner of neuroplasticity include synaptogenesis and synaptic pruning. This is the effect of creating or removing the whole synapse, adding or taking away connections between neurons. Neurogenesis is the creation of whole new neurons, and neural cell death is when neurons die (this can be a natural process, but could also be as a result of damage, overexcitation or disease). These usually come from brain scans and can involve changes in the density of white matter or grey matter on structural MRI scans, or how densely radioactively labelled markers bind to specific receptors in parts of the brain.

There are other forms of neuroplasticity, including when we see from brain scans that areas of white or grey matter have become more dense, but we don't actually measure what individual neurons are doing. We look at examples of this in Chapter 28 on mindfulness.

As is common with things relating to the brain, we learn a lot when things go wrong. Brain injury has given us a good insight into the necessity for neuroplasticity. If an area that we believed was solely responsible for movement of the left arm, for example, was destroyed in an accident, then the old way of thinking would suggest that the person would not be able to move this arm again. However, with neuroplasticity it is possible for cortical remapping to occur, so for the role of moving the left arm to be taken over by other neurons in a different place in the brain.

This isn't the only time that the brain changes. In fact it turns out to be a normal process that is happening regularly. Behaviour changes, environmental changes and thoughts can all effect change in neural pathways and synapses. You'll remember when we looked at the synaptic gap and neuro-transmitters binding to the receptors on the postsynaptic cell that either an action potential was generated or it wasn't. The additional piece of information here is that the *speed* and therefore the *frequency* of the action potentials can be influenced. Typically, the more this particular pathway is activated, the faster and so more frequently a nerve impulse can be seen to be travelling.

A famous paragraph written by Donald Hebb in 1949 reads:

> Let us assume that the persistence or repetition of a reverberatory activity (or 'trace') tends to induce lasting cellular changes that add to its stability... When an axon of cell A is near enough to excite a cell B and repeatedly or persistently takes part in firing it, some growth process or metabolic change takes place in one or both cells such that A's efficiency, as one of the cells firing B, is increased.
>
> (Hebb, 2002)

Hebb's concept is often paraphrased to 'Neurons that fire together, wire to-gether'. In fact, it was a Spanish neuroanatomist in 1894, Santiago Ramon y Cajal, who first proposed a way that we could learn that wouldn't involve the creation of new neurons. The exciting thing is that we have the power to influence the very structure of our brains. For coaches this is key. This could well be the most important chapter in this book, and content that should be included in any book on coaching! Certainly every coach should have a thorough understanding of this topic.

Experience produces changes in the brain, some of the changes that occur being:

- increased dendritic length;
- increased dendritic spine density (or decreased);
- synapse formation;
- increased glial activity (commonly considered as the glue of the nervous system);
- altered metabolic activity.

There is not scope here to go into all the research that suggests the variety of options open to coaches, leaders and managers to help encourage neuro-plasticity. However, we will touch back on a selection of strategies in later chapters.

Why is it important to me as a coach?

Neuroplasticity takes the centre stage here in the relatively recently recognized ability of the brain to change. From the micro changes in neural pathways right up to the macro effects typically accompanying a brain injury we are now convinced that the brain can physically change. This fundamental process is at the heart of most of the work that coaches do with their clients. In fact, we believe that coaching facilitates 'self-directed neuroplasticity'.

When we share that belief during the Neuroscience for Coaches training programme most coaches get quite excited because they see the power of neuroplasticity. It underpins the basis of change work.

There are many old and fascinating studies in this area conducted by people such as Paul Bach-y-Rita and Michael Merzenich. Some of the early research involved things such as helping blind people to read and perceive shadows – this was dubbed 'the first and boldest applications of neuroplasticity'. It involved an electrically stimulated chair with a large camera that sent signals of what the camera saw to stimulators against the subject's skin. The results were as profound as the individuals eventually being able to recognize a picture of Twiggy! The profoundness of these results is hard to comprehend. People who could not see were able to process visual information – the brain behaved with a huge amount of plasticity.

We explore several studies around neuroplasticity in *Make Your Brain Work* (2013), including some of the work by Michael Merzenich. He is very bold in his understanding of what neuroplasticity can accomplish. He suggests that brain exercises may be as useful as drugs to treat diseases as severe as schizophrenia.

An experiment was done that focused on the plasticity of working memory (Li *et al*, 2008). The ability to hold things in our working memory for short times is a very valuable one. The study looked at what effect extensive working memory practice had on performance improvement, transfer and short-term maintenance of practice gains and transfer effects. Two age groups were included, 20- to 30-year-olds and 70- to 80-year-olds. They all practised a spatial working memory task with two levels of processing demands for 15 minutes per day for 45 days. The results showed substantial performance gains (so they got a lot better), near transfer to a more demanding spatial and numerical task (so they were able to transfer their skills to harder tasks) and three-month maintenance of practice gains and near transfer effects (so the results lasted).

A study that as an ex-medical student I find amusing, as well as fascinating (and experientially believable), is one that focused on the brains of medical students (Draganski *et al*, 2006). These volunteers had their brains scanned before and after studying for their exams. In this short period of a few months their grey matter increased significantly in the posterior and lateral parietal cortex.

A case that may be motivating to many people is that of Rudiger Gamm, a German who became a mathematical phenomenon. Born with no special powers, he can now calculate the ninth power or fifth root of numbers and solve problems such as 68×75 in five seconds! From age 20 he spent four hours per day doing calculations, resulting in him at the age of 26 being able to earn a living performing such calculations on television. Positron emission tomography (PET) scans revealed that he was using five more brain areas than was the norm for these calculations.

Neuroplasticity underpins learning, remembering things and changing behaviours. Experiences can change both the physical structure of the brain and the functional organization of it. This in turn affects how we respond to experiences. One of the most well-known examples of this is the practice of meditation, which has been shown to increase the actual physical size of different areas of the brain. We will look at this more in Chapter 28 on mindfulness.

Some things you could try with your clients include eliciting something they would like to work on, for example being more patient with their children, then focus for two months on strengthening the neural networks associated with patience. This might include visualizing situations where previously they would have become frustrated but instead respond in a calm manner, taking a deep breath before they respond to their children, doing things that they know will require patience and talking to people about successes they've had with being patient. There is good reason to let your clients 'sleep on it'. When we sleep we consolidate learning and memory. Sleep is also a great time for plastic change to occur in the brain.

Threat response

17

What is it?

The threat response is a series of reactions that occur within the body when we *perceive* a threat – the same as if there were a real threat present. We use the term 'threat' to include a range of subtle triggers, which have not all been conclusively researched. When I speak to groups of leaders there can be confusion between this and the fight-or-flight response because the basic physiological response is similar; however, the physical effects of the threat response are normally less stark, while the mental ones remain potent.

Why is it important to me as a coach?

The threat response underpins so much of a client's day, unconsciously, that it can drive a lot of their actions. Understanding what some of the triggers are and how they work enables you to ask better-quality questions and direct attention more effectively. There are many studies that look at the effects of various triggers of the threat response that we cover in various other sections of this book. By making your client aware of potential triggers, and how triggers may compromise others, you are enabling them to be more effective in a huge range of situations.

For managers in a coaching role it is especially important to realize that we know that when the threat response is activated the body is flooded with noradrenaline, which can be useful for increasing focus. However, the whole stress response overall isn't great because it:

- inhibits creative thinking;
- tires the brain quickly, reducing cognitive resources;
- can have negative health impacts due to increased cortisol;
- impacts on your ability to think;
- can make you behave or think more defensively;
- can make you less able to process new information.

The threat response is very powerful and can be long lasting. Much more so than the 'toward' response you get with a potential reward, which is sometimes disappointingly short.

Ultimately these responses are hugely important because they have a direct effect on the efficiency, effectiveness and productivity of people. For example, if an employee feels they are being micromanaged by their leader it can directly affect their productivity, which ultimately costs the company money. So coaching leaders to effectively work with people and avoid triggering the threat response is vital.

The neuroscience of the threat response

The threat response itself has not been thoroughly researched in isolation. Rather it can be thought of as a collective term for the bringing together of lots of research that identifies a below optimal performance from individuals. As we have noted, some of the physiological responses are similar to those of the fight-or-flight response.

The neuroscience of the fight-or-flight response is well documented. The autonomic nervous system is made up of both the sympathetic and parasympathetic systems. It is regulating many unconscious processes every minute of every day. For example, it controls our heart rate, digestion and breathing rate. The sympathetic nervous system activates the release of norepinephrine when a threat is presented. The parasympathetic system can be thought of as the restorative system releasing acetylcholine and bringing back a homeostatic balance.

In the fight-or-flight response the amygdala triggers a neural response in the hypothalamus. The pituitary gland is then activated and secretes adrenocorticotropic hormone. Around the same time, the adrenal gland is activated and releases epinephrine. Cortisol is produced and this has various effects, including suppressing the immune system, increasing blood pressure and increasing blood sugar levels. This creates an energy boost, along with an increase in heart rate and breathing rate, preparing the body for fight or flight.

Turning our attention now to the threat response. When a person experiences a threat response a series of events inside their body are triggered. The process itself uses up oxygen and glucose. This has the effect of decreasing working memory capacity. You need your working memory for creative insight, analytic thinking, problem solving and even just simply holding something in your very short-term memory. These are all impaired when you have a threat response.

A threat response also affects the amygdala, anterior cingulate cortex and frontal lobe. The release of cortisol is triggered and this results in decreasing immunity, impairing learning and affecting memory. Together, this spells disaster for efficiency, effectiveness and productivity.

The brain cannot distinguish between real and unreal threats. For example, if you see what may be a snake near you on a country walk it is safer to respond as if it is one, rather than presume it isn't. Your limbic system responds quickly to keep you safe. If it turns out that there isn't a snake, your brain usually calms down the threat response. In organizations things are normally far more complicated and your brain often struggles to determine if something is a real threat, so it keeps you safe by presuming it is!

When a threat is perceived, just like in the fight-or-flight response, the important system called the hypothalamic-pituitary-adrenal (HPA) axis kicks into action. This system comprises interactions between the three named brain areas. The HPA system triggers the production and release of cortisol and epinephrine. This prepares the body for action; the blood pressure and heart rate increase, the lungs take in more oxygen by increasing the breathing rate and blood flow can dramatically increase.

Neurotransmitters are also released as a result of the HPA system's activation. These chemical messengers activate the amygdala, which then triggers the brain's response to emotions. The chemical messengers also tell the hippocampus to create a record of this emotional experience in the long-term memory. Finally, the neurotransmitters also suppress activity in the frontal lobe, which means that short-term memory, concentration, inhibition, and rational thinking all take a nosedive. Trying to handle social interactions elegantly or do cognitive tasks becomes very challenging while in this state.

Suggestions

The most important thing with the threat response is to educate people that it exists. Once people are aware that it can be hampering their efforts in many different areas of their life they are then in a position to do something about it. In the workplace imagine the difference if all leaders, managers and team members knew that this was a possibility for people they were dealing with. The opportunities for increased empathy, patience and understanding would grow.

Potential triggers of a threat response include:

- a perceived lack of control;
- feelings of uncertainty;
- any sense of unfairness;
- poor connectedness.

So both exploring where these may be occurring in your client's life and also what may be going on for other people in their lives at certain times – friends, colleagues and family – could be useful. It could also be worth identifying ways to strengthen any recurring areas of susceptibility. For example, if someone often experienced a threat response at work due to a perceived lack of control there might be steps that could be taken to reduce the frequency of this. Is it because their boss has a tendency to micromanage them? (If so, could they have a conversation with the boss?) Is it because customers keep changing their mind about what they want? (If so, could procedures be put in place to minimize this?) Or is it because they are filtering for the things that they cannot control? (If so, would some work on focus and filters help?)

Neuroimaging

18

What is it?

The field of neuroimaging is a specialized one. It includes all the techniques currently at our disposal to study the structure and functions of the brain. These two main categories, structure and function, enable us to focus in on different things. Structural imaging looks at brain anatomy and is used in a medical context for diagnosing things such as lesions or disease. Functional imaging looks at brain activity and so it is used for the cognitive psychology research that you will read a lot about in this book. We talk about brain areas 'lighting up', this is because there is an increase in metabolism in a specific area and it then shows up with definition on functional imaging scans – we then know that this area of the brain is processing information.

So what are some of the research tools that neuroscientists use?

By purpose of a historical overview we'll start with the computerized axial tomography (CAT or CT scanners). These were introduced in the 1970s and gave us our first, now familiar, pictures of the inside of the body, which were useful diagnostically and for research intent. The inventors, Allan Cormack and Godfrey Hounsfield, received the 1979 Nobel Prize for Physiology or Medicine for this and it was a real boost to neuroscience. Shortly afterwards, the single photon emission computed tomography (SPECT) scan and positron emission tomography (PET) scan were facilitated by the development of radioligands. This was in the early 1980s, when the magnetic resonance imaging scan was also being developed. Its use initially was mainly clinical. Functional magnetic resonance imaging followed shortly after this and, in 2003, Peter Mansfield and Paul Lauterbur were awarded the Nobel Prize for Physiology or Medicine for the development of the MRI scanner.

Computerized axial tomography (CAT or CT)

This form of scanning uses X-rays. We see an end result of pictures that look like many cross-sections of the brain. The bit in between involves a computer program that performs a numerical integral calculation estimating how much X-ray beam has been absorbed. The denser a part of the brain is, the whiter it will appear. This is just like 'normal' X-rays where bones show up as a strong white and we can then easily (well, easy-ish) see where there are fractures because they show up as dark lines. CT scans have most value when a picture of the brain is needed quickly because there are concerns about swelling or bleeding. You can get good images quickly with a CT scan, whereas with fMRI the process takes a lot longer. CT is accurate, reliable, safe and so really good for people who have experienced a head trauma.

Positron emission tomography (PET)

This involves the use of a radioactive chemical (tracer) that collects in cells that are using a lot of energy and a special camera that can record the positively charged particles that the tracer gives off. Individuals are injected with this compound and it travels through their bloodstream up to the brain. The resulting scan is a multicoloured two- or three-dimensional image that highlights where the compound acts in the brain. Different compounds can show blood flow and oxygen and glucose metabolism in the tissues. The scans are quick, indeed they can only measure short tasks as the radioactivity decays quickly.

The picture isn't as detailed as MRI or CT scans and tends to have clinical uses, such as looking for cancer, Alzheimer's disease and strokes. These all cause observable changes in metabolism so are easily picked up by PET scans. PET scans are able to identify specific receptors associated with particular neurotransmitters. This is a unique and valuable feature.

Single photon emission computed tomography (SPECT)

This also involves administering a compound to the person, in this case a gamma-emitting radioisotope. We also get a two- or three-dimensional image of the brain regions that are active. This type of scan is very useful for people with epilepsy, because once you inject the radioactive tracer it is distributed within the brain 30–60 seconds later. You can then see what the cerebral blood flow was when the tracer was injected. The downside is that the scan has poor resolution.

Magnetoencephalography (MEG)

This measures the magnetic fields produced by electrical activity in the brain. MEG delivers a direct measurement of neural electrical activity. The temporal resolution is high and the spatial resolution is low. The results from MEG are considered less distorted than from electroencephalography (EEG). MEG is used for research, in surgery to localize pathology and for neurofeedback, among other uses.

Functional magnetic resonance imaging (fMRI)

Functional magnetic resonance imagining uses magnetic fields. Due to the use of magnetic fields there are some people who cannot go into these scanners. People with cochlear implants, pacemakers or other surgical implants, some forms of tattoos and people at risk of having shrapnel in them.

It measures the proportion of deoxyhaemoglobin in the blood, which is correlated with brain activity. The paramagnetic properties of oxygenated and deoxygenated haemoglobin are key to how we are able to see the change in activity in the different areas of the brain. Resolution is 2–3 millimetres, which is great. People can be asked to do simple tasks while in fMRI scanners, for example pressing a button; this has dramatically increased what we can do from a research perspective. For coaches, the fMRI scanner is used most in the types of studies we are interested in. Most of the studies you will come across (certainly the ones we tend to refer to) utilize fMRI scanners.

Why is it important to me as a coach?

The difference between a coach who has a depth of understanding versus one who has only a surface-level understanding is significant. There are many benefits to really getting to grips with this information, and a foundational piece of knowledge is about knowing how data is gathered. It is similar to going into an organization and being given a report about someone. You would rightly want to understand how that information had come to be on the report. Did the individual self-report after some reflection? Was the report the result of a 360-degree evaluative process? Was the report written by the individual's boss? Or was it compiled from an exit interview from a disgruntled previous team member? As you can see, where the information came from can dramatically affect the filters with which you read the report.

If you choose to give yourself a thorough grounding in the foundations of how the brain works by attending a programme on Neuroscience for Coaches then (at least from our programme) you will be equipping yourself to read scientific papers in the future. Knowing how the equipment that helps gather the data actually works gives you an insight into the scope of interpretation from that data.

Working memory

What is it?

'Working memory' is a term that psychologists came up with to describe our ability to hold things in the front of our mind and manipulate these small chunks of information. A classic example is mental arithmetic. The term is often used synonymously with short-term memory, and is thought by many to supersede the older 'short-term memory' term. When a task is performed using the working memory system it involves executive and attention control enabling integration, processing, disposal and retrieval of information. It is important to remember that this is a concept that at this point is theoretical, a working model that is useful to our understanding of things.

We believe that frontal cortex (especially the dorsolateral prefrontal cortex), parietal cortex, anterior cingulate and parts of the basal ganglia are important in short-term or working memory. The term 'working memory' was originally coined in 1960 by George Miller, Eugene Galanter and Karl Pribram and was linked to the very computer-like model of the mind. When we talk of using our working memory we are talking about a couple of seconds; it has a limited capacity. In 1956 Miller proposed that we could hold seven units in our mind. In neurolinguistic programming (NLP) this is referenced as the 'magic 7 plus or minus 2' chunks (Miller, 1956). We know now that there are several factors that affect how many chunks a person can hold in their working memory. For words, for example, this is dependent on the length of the words and the familiarity of the words.

Why is it important to me as a coach?

Clients use working memory very frequently. There are ways to engage with it to help optimize its performance, and things that are unhelpful when working with it. Being aware of what it is and how it works means you can help your clients notice strategic changes they could make. Perhaps you might even want to invite clients to do certain things in order to improve their working memories.

As you have just read, the 'people can remember 7+ or −2 chunks' has since been revised. Various factors are involved in how many 'chunks' a person can hold in their working memory. Something to bear in mind when we are discussing working memory is the research of K Anders Ericsson, who has studied individuals who have been able to remember up to 80 chunks (Ericsson, 1980). This, and similar memory experiments, can be quite deceptive though as this individual had in his long-term memory lots of digits from other reference points, in this case racing times. Ultimately at the working memory level there was a relatively small number of chunks being stored, but each of these unpacked a lot more digits below. Almost like a code or a parcel that unwraps (or act as retrieval cues) and gives a lot more data at the next level. Using memory-training skills therefore would not increase working memory. Indeed this individual with a great digit memory had normal word-recall ability.

A randomized controlled study showed that working memory training increased a range of cognitive abilities and that brain activity in the prefrontal cortex increased. Also, measurable increases in the density of dopamine neuroreceptors were noted. From the neurotransmitter perspective both dopamine and glutamate are thought to be involved in working memory.

A key study, for example, highlighted how important our ability to suppress irrelevant information is for our working memory to work optimally. Distractions need suppression, which often means that the neural activity in the sensory cortex needs modulating. In very simple terms, our ability to remember relevant information depends on our ability to limit interference from irrelevant information. This hopefully sounds very logical and resonates with you. Until studies are done to confirm the mechanisms that are occurring at a neural level we don't know some of these things for sure, we are only guessing that what sounds logical on an experiential level could be mimicking what is occurring at a brain cell level. It has now been shown that when we are distracted by irrelevant information the neural process involves an increased memory load. Our ability to effectively filter irrelevant information by neural suppression, in order to prevent overloading our limited working memory capacity, is a key process.

Coaching provides a wonderful opportunity for self-evaluation. Inviting a client to give themselves an honest appraisal of where they feel their current ability to suppress irrelevant information lies could be a very valuable experience. Considering different areas of life – for example, at work when on our own, at work when with others, at home on our own, at home with partner, with children, etc – could bring to light some opportunities for development. Of course there are environmental things we can do to help reduce the need for the suppression of external distractions. One of the easiest at work or home would be to tidy up. Clear out all old stuff that you are not currently using. Put things out of sight that are not in regular use.

So we know that our ability to store things in our working memory and process information there is impaired when we allow distractions to do their job. The question could then become: what can we do about that? The answer lies partly in a study that looked at practice-related improvements in working memory. This study used electroencephalographic (EEG) recordings with individuals while they performed three motion-direction, delayed-recognition tasks. The first task didn't have any interference associated with it, the other two involved different types of interference during the interval of memory maintenance: distractors and interruptors. Both of these forms of interference disrupted working memory performance. However, both accuracy and response time improved with practice. Thus it would appear that we could get better at filtering irrelevant information.

One favourite study looked at the effect of chewing on working memory (Hirano *et al*, 2008). Chewing has been linked to the activation of various brain areas, including the prefrontal cortex. This study looked into the specific effects of chewing on neuronal activities during a working memory task and used fMRI to help see what was happening. The participants of the study had to chew odourless and tasteless chewing gum and perform specific tasks of working memory. The results were complex, showing activation in lots of areas of the brain, including the dorsolateral prefrontal cortex. There was also the suggestion that chewing may accelerate the process of working memory.

For anyone working in a knowledge-based role, where their ability to think or to compute are important assets to them, there is room to support growth. Consider teaching your client what working memory is and what its limitations are. Remember:

- When we get distracted we often lose the contents of our working memory.

- If someone is distracted when giving us important information then the information is unlikely to even get into the working memory – so it hasn't even been properly heard... let alone registered or remembered.

- Your working memory has a limited capacity – so expecting it to hold lots of information is futile. Also trying to do complex things, such as 279×534, for most of us would be tricky because of the number of steps and different chunks of information that need to be stored.

- Everyone is different – sometimes this may be down to differing working memory capacities, but sometimes, like in chess or mental arithmetic, it is also down to a different type of memory being utilized.

- High-intensity exercise has been shown to increase the capacity of working memory.

- Stress impairs working memory – so if your client is stressed they can't expect to have their working memory in optimal working condition.

Overall, it is suggested that by improving our working memory we can become happier, more focused and more effective. Some odd ways you could improve it include:

- Climbing a tree – you'll use your working memory to balance and plan your route while evaluating your current position.
- Eat some red meat or oily fish or dark chocolate (temporary boost).
- (The big one) turn off your smartphone – the constant interruptions are deadly to prolonged focus and getting the most out of your working memory.

Multitasking has been correlated to the processing and storage components of working memory capacity. The more effective a person's working memory is, the better at multitasking they appear to be. This is especially useful in certain job roles. We know that practice can have a big effect on the brain, so investing time in becoming really good at focusing on one task at a time could increase the synaptic connections in areas of the brain involved in working memory.

HPA axis

What is it?

The hypothalamic-pituitary-adrenal axis is part of the neuroendocrine system that is involved in our reactions to stress. It is also important in regulating many other processes such as digestion, the immune system, mood and emotions, energy storage and expenditure. Essentially it is made up of a series of detailed interactions between the hypothalamus, the pituitary gland and the adrenal glands.

The hypothalamus is responsible for making and releasing vasopressin and corticotropin-releasing hormone (CRH). Together they regulate the pituitary gland. Adrenocorticotropic hormone is released, which acts on the adrenal cortex. This produces cortisol, among other things. They then act on the hypothalamus and pituitary in a negative feedback cycle.

CRH is released from the hypothalamus in response to several things including stress, physical activity and illness. It is also linked to the circadian rhythm of an individual. You'll remember from Chapter 9 on cortisol that levels rise rapidly after we wake and reach a peak after 30–45 minutes. Abnormalities in this cycle are linked to insomnia and burnout.

Connections between the amygdala and the hypothalamus mean that sensory information that arrives at the amygdala can be communicated to the hypothalamus quickly. So if we see, hear or in any way perceive something that is interpreted to generate a fear response then the HPA can be activated, along with the sympathetic nervous system. As we've discussed previously, in small doses the release of glycocorticoids (of which cortisol is one) is important, but released in excess damage can be done. The hippocampus can be affected, so memory can be impacted.

Why it is important to me as a coach?

HPA is involved in many disorders, including anxiety disorder, burnout, irritable bowel syndrome, alcoholism and insomnia (also many others but it is unlikely that you would be working with people suffering from many of them). The overactivity of the HPA is also said to be a characteristic of depression. Antidepressants act to regulate the HPA axis function.

Neurotransmitters such as dopamine, serotonin and noradrenaline have a role to play in the regulation of the HPA axis. Studies show that positive social interactions that increase levels of oxytocin suppress the HPA axis and this can have benefits that include improved wound healing. Being aware of the positive and protective impact that oxytocin can have is important when working with individuals who are at risk of chronic stress.

One study addressed the question 'Do burned out and work-engaged employees differ in the functioning of the hypothalamic-pituitary-adrenal axis?' (Langelaan *et al*, 2006). The saliva of just under 100 managers was collected on several days and the levels of cortisol and a compound that counterbalanced cortisol were measured. The group that were work-engaged were found to have stronger cortisol suppression in response to the other compound, which suggested that they have higher feedback sensitivity. So work-engaged employees do differ slightly in the functioning of the HPA axis.

Mirror neurons

What are they?

The mirror neuron is a type of neuron that fires when an action is observed in another neuron. The story of how mirror neurons were initially discovered is an interesting one (*New York Times*, 2006). In the 1990s in a lab in Parma, Italy a group of neuroscientists were looking at the neurons of monkeys. The monkeys were wired up to a machine that made a noise when they reached for/ate a peanut. One day, after a graduate student came back from lunch (still eating his ice cream) the machine made the noise... but the monkey hadn't moved. This was strange because the expectation was that the neurons only fired and triggered the machine when the monkey moved to eat something. What they found was that the neuron fired when it observed others eating, as well as when the monkey ate. We now know that this occurs in humans too. Initially when the researchers tried to report on their observations their first paper was rejected. This is one of the great, and challenging, things about scientific fields such as neuroscience. It is often only with the building of information that new ideas become accepted. Currently there is still speculation around this system and it can be useful to think of mirror neurons as part of the action-observation network.

The action-observation network (AON) has been suggested to support our understanding of others' actions and goals. People can track and in part replicate the mental and bodily states of others. Neuroimaging studies have shown overlapping activation in the brain for perception and execution of motor actions in a parietofrontal network involving several brain areas. The AON is suggested to support social perception by interacting with the somatosensory cortices.

A study led by Lauri Nummenmaa published in 2014 (Nummenmaa *et al*, 2014) involved subjects watching movies showing boxing matches while in an MRI scanner. They either watched the movie passively or simulated the feelings of one of the boxers. When simulating the feelings, multiple brain regions were activated and the interconnectivity of these regions strengthened during the process. The researchers proposed that sharing a third person's feelings synchronizes the observer's own brain mechanism's supporting sensations and motor planning. This makes mutual understanding more likely.

It is suggested that similar brain states could be a prerequisite for similar mind states. This makes logical sense. Other studies have shown overlapping neural activation in perception, experience of emotional states such as pain, disgust and pleasure.

For many years it was thought that mirror neurons were only related to motor functions. Now it is thought that, in Dr Rizzolatti's words: 'Mirror neurons allow us to grasp the minds of others not through conceptual reasoning but through direct simulation. By feeling, not by thinking' (*New York Times*, 2006). These special mirror neurons have been found in the premotor cortex, posterior parietal lobe, superior temporal sulcus and insula. When we are in a meeting and someone is struggling, either forgetting things or having to share bad news, people often 'tune in' to what is going on for that individual. They feel what that person is feeling. Some neuroscientists say that this is down to mirror neurons.

In recent years some neuroscientists have gone a step further than thinking that mirror neurons only have a role in mirroring motor actions, and have suggested that mirror neurons are also important in:

- theory of mind skills;
- helping us to understand the actions and intentions of others;
- emotions such as empathy.

These obviously are involved in a lot of our daily lives and so awareness of a potential influencer is important. If mirror neurons do help us to understand the intentions of other people then this is another reason to support clients in trusting their instincts. Strengthening trust in oneself is also important if mirror neurons can pick up on and help us to perceive the feelings of another person and so enable us to be more empathetic. Right now, however, the official party line (if one exists) is that there is not a widely accepted neural model to explain how these things may work. Yet that does not mean that in time there won't be such a model.

A study by Christian Keysers (2011) found that when people watch a hand extend to caress someone but be intercepted by another hand pushing it away (rudely) then the insula of the observer registered the social pain of rejection. It should be noted that the existence/extent of mirror neurons outside of motor neurons is strongly debated. Studies are cropping up every year and the neuroscientists are thrashing out their differences as you read this!

Giacomo Rizzolatti, Christian Keysers and others have been investigating the role of the anterior insula and anterior cingulate cortex in order to better comprehend how we understand the emotions of others. In one study the researchers looked at the fMRI scans of participants while they inhaled odorants that produced feelings of disgust in them (Keysers *et al*, 2003). They also then watched video clips of people showing the facial expression of disgust. The results showed that both seeing disgust in others and feeling disgust oneself activates the same sites in the anterior insula and to a lesser degree the anterior cingulate cortex. Keysers says that there is a uniquely human mirror-neuron system found in the insula. He also says that people who are highly empathetic (as ranked on a scale measuring empathy) have particularly active mirror neuron systems.

Why are they important to me as a coach?

Consider this example: there has been a stressful meeting and everyone is tense. Bad news has been delivered to the people around the table. Before anyone realizes it, people have started to attack each other. Your client focuses on exploring the surface-level challenges, thinking Jo is wrong or Bob needs to behave differently. The surface-level challenges are likely not the issue, however. It is possible that mirror neurons are creating an intense group of stressed individuals, the core cause being fear rather than any surface problem. As one person responds in an aggravated fashion, others do likewise. This isn't necessarily down to mirror neurons at all, although it just might be possible.

There are also many potential implications for these mirror neurons in many areas of business. One of the most neglected among coaches overall is that of marketing. Understanding that by seeing someone do something mirror neurons are fired off in the observer means that showing someone something is hugely powerful. For example, on your website do you have pictures of people enjoying the coaching experience? What about videos of people taking the type of action that your coaching supports people to take?

When you are working with a client could you conduct a session focusing on the external influences that may be affecting them? Do they read the newspaper each morning, filling their mind with pictures of disturbing things that may be putting them in a less than optimal working state... is there a better time to read that newspaper? We know that repeatedly activating areas of the brain strengthens neural connections and can increase the size of different brain regions... so what are they frequently looking at?

A caveat to each of the above areas for consideration is that mirror neurons work best in real life. So perhaps the opportunity for face-to-face coaching sessions should be taken when it can be to maximize motivation, inspiration and positivity. Could you even give people, perhaps, the opportunity to experience watching a coaching session?

PART FOUR
Brain networks

Many of us like things to be simple. We like to be able to break things down into chunks that are easy to understand. We have started with that approach to the brain, but it isn't the whole story. While the brain has fixed anatomy (as opposed to the heart with moving parts) it has many functions (the heart has one function). The brain is very complex. This complexity of connectivity of the network architecture enables us to respond to the range of environmental demands we are subjected to.

Brain networks

22

Looking at the brain from the position of its networks is a different approach and one that offers another level of understanding. Our comprehension is expanding in neuroscience at a rapid rate. Concepts that were once favoured – some for their ease of understanding and applying, for example, the idea of people being 'left or right brained' – perhaps didn't have solid research underpinning them and are falling away. We are left with the myths, though. For that reason, our approach is one that prefers to clearly identify the science.

Hot and cool

In psychology a dual system is quite widely explored, with the foundations of the thinking being attributed to William James. Many theories have followed, among them Daniel Kahneman's interpretation in 2003 – of which a *Horizon* television documentary in February 2014, presented by Kahneman, has further raised the profile – of this hot and cool 'system 1' and 'system 2' way of processing information.

System 1 is implicit (automatic) and unconscious, whereas system 2 is explicit (controlled) and conscious. You are probably familiar with the classic experiment where children are asked to sit with a treat such as a marshmallow in front of them and decide whether to eat it now or wait for 15 minutes to receive another one. The researchers suggested that this was an example of the conflict between the two systems and that most children are challenged because their cool system is underdeveloped.

From a neuroscientific perspective we can use MRI to look for the possible underpinnings of these networks in order to add another layer to our understanding. In a study led by McClure (McClure *et al*, 2004) functional MRI was used to observe brain region activation while the participants in the study made decisions about monetary rewards. The results showed that when choices involved immediate gratification, areas of the limbic system were activated. You might remember the limbic system as being considered one of the 'old' parts of the brain from an evolutionary perspective. It is involved in emotion, behaviour, motivation and long-term memory. When participants were making choices that involved delaying gratification, the cortex was more active. This is the evolutionary newer area of the brain that forms the outer layer of its anatomy. The regions here are important for cognition.

We can think of system 1 (or the 'hot' system) as being emotional, unconscious, automatic (so fast), simple, evolutionarily old and nonverbal. Whereas system 2 (or the 'cool' system) is cognitive, complex, self-controlled, slower, involves effort, logical and is limited by our working memory capacity.

This network is especially important when we are considering:

- Choice architecture – which can utilize the hot network to get immediate results.
- Habits – where we are often operating in system 1.
- Decision making – where system 1 could be playing a bigger part than we realize.
- Expectations – the dance that occurs beneath system 2's radar.

Default mode

The default mode network is the collection of the areas of the brain that increase in activity when we are not focused on a task. It is sometimes referred to as a resting state network. The areas include the medial prefrontal cortex, medial parietal cortex, posterior cingulate cortex and medial temporal lobes. The research is in its early stages and so some of the consequences of these cortical associations have yet to be uncovered.

When we have thoughts that are independent from our immediate environment the default mode network appears to be active. Examples include when our mind is wandering, when we are being introspective or when we plan what we're doing some time in the future.

One of the theories about default network activity is that it is the brain's baseline of processing and information maintenance. The suggestion is that the network helps us consolidate our experiences and prepare to effectively react. Interesting people with specific medical problems such as attention deficit hyperactivity disorder (ADHD), autism, Alzheimer's and schizophrenia have been seen to have different activity in this network. Additionally, people who have experienced long-term trauma, such as abuse, appear to have lower connectivity across the default network.

Dorsal/ventral attention stream

The final network we will look at here is a model for the neural processing of vision. As with many models it is a work in progress; the components are useful to us now, and further research may develop the model. It suggests that after visual information has gone into the occipital lobe (the visual-processing lobe of the brain, found at the back of the brain and home to most of the visual cortex) the information then follows one of two main streams.

The ventral stream is also known as the 'what pathway'. It is involved in us being able to identify and recognize objects and subsequently plan behaviour. The other stream is known as the 'where pathway' or the dorsal stream. It is involved in processing the location of objects relevant to us in order to programme our behaviour. The ventral stream travels to the temporal lobe, while the dorsal stream heads to the parietal lobe (two of the four lobes that the brain is divided into, the others being the occipital and the frontal lobes).

So we have one pathway providing a detailed representation of the world for cognitive processes. It is relatively slow, involves long-term stored memories and there is typically a high level of consciousness around it. The other pathway is translating the visual information into coordinates required for skilled motor behaviour – it helps guide us. It is relatively fast, involves only very short-term memory storage and typically involves low consciousness.

These streams are a beautiful illustration of how brain areas work together. Remember that the brain is not designed as separate components. We observed it, cut it up and labelled different sections. When we see the passage of visual information from our eyes into our occipital lobe, where visual functions are happening to the spatial awareness functions in the parietal lobe (as is the case with the dorsal stream), we get a real feel for the networks that are present in the brain. Considering that the areas in the ventral stream are influenced by factors such as attention, working memory and the salience of stimuli, we can see that its role is also in evaluating the significance of what we see.

Network conclusion

The real opportunity available to us after looking at some brain networks is to be open to this different way of looking at the brain. The temptation is to focus on 'what bit of the brain does this?' when rarely is it that simple. Just like in organizations where several departments often talk to one another, work on the same project or socialize together, the brain is also very interconnected.

PART FIVE
The quantum brain

This is a very special part of the book. Often when working with organizations it is interesting to note that in all types, from large corporate ones right through to smaller entrepreneurial ones, individuals get very excited when quantum physics is mentioned. Originally this section slotted into Part Three of the book, however, after reflection it got its own section. There are a couple of reasons for this. One, the content of this section is quite controversial. Speak to many physicists and they will tell you that several things mentioned are not possible. Speak to many neuroscientists and they will tell you that these lines of thought are not going anywhere. However, if there is one thing that science has taught us down the ages, it is that we can be wrong. So in the next 10–25 years one of two things is likely to happen: either this section will be removed from future editions due to additional research rendering it unworthy, or it will link to a whole new book, which expands what we currently think. Personally, I believe it will be the latter, which is why it is included here.

The second reason that the quantum brain got its own section is that there is a lot covered here and before reading it you will want to ensure you have the head space to dive into all the deep content. So take the opportunity to refresh yourself and get ready for some information that could change everything.

The quantum brain

23

What is the quantum brain?

The prevailing understanding of how the brain works, indeed how most of the world works, is in alignment with classical physics. However, there are some people who believe that what we know about more contemporary physics, namely quantum physics, is also important in our understanding of the world. For most people this is an alien concept and the implications seem quite 'out there' or unbelievable. A traditional and pure neuroscience approach tells us that the brain is made up of material particles and fields, and that everything can be explained in these terms.

Quantum physics challenges these ways of thinking. It offers experiments that show that human thoughts, choices, will is important. If, as some scientists advocate, this key concept is applicable to neuroscience then it gives us an entirely fresh conceptual framework. As with many things historically in science, good ideas often take a little time to move from the sideline into the mainstream. Some of the strong drivers pushing these understandings forward are as follows. The structural features of ion channels that are fundamental to synaptic function need quantum physics to analyse them. Quantum physics is also more appropriate in neuroplastic mechanisms that involve directed attention and mental effort.

Why is it important to me as a coach?

Contemporary physics is of importance to everyone because our under-standing of our world impacts our beliefs and our values. In classical physics we can reduce everything to a mechanical process. In quantum physics the causal role of human consciousness is important in our reality. Niels Bohn, Werner Heisenberg and Wolfgang Pauli had intense exploratory discussions, the result of which was valuable new ways of thinking. In these new ways of thinking our value-based intentional efforts can make a difference in our own behaviours, our social environment and our larger reality.

In classical physics you can understand a system at a single point in time by knowing the location and velocity of all the particles within that system. You also need to know any fields that are acting on them (such as gravitational and electromagnetic). Any human's consciousness is not considered important in classical physics from the perspective of being involved in any way. Classical physics is deterministic whereas quantum physics is probabilistic.

Henry Stapp, a theoretical physicist, summarizes that in psychology what people allow into their consciousness is strongly linked to what their mind intends; people experience what they are looking for. Providing the potentiality for an experience to occur, the will of the individual draws out this from the range of options. Stapp likens this to quantum theory. He and two colleagues explore this further: Schwartz, Stapp and Beauregard (2005) suggest there are five facts that mean that it is advantageous to neuroscience to use the frameworks from contemporary physics rather than classic physics:

1   They suggest that the terms 'feeling', 'knowing' and 'effort' are intrinsically mentalistic and experiential and so cannot be described exclusively in terms of their material make-up.

2   In order to explain how big systems work, the founders of contemporary physics had to recognize the importance of choices made by humans about how they will act on the behaviour of the small components (the atoms) of these big systems.

3   This new conceptual framework has choices described in psychological language.

4   This terminology is very important to design and execute the type of experiments that demonstrate self-directed neuroplasticity.

5   What was done in atomic science could be also done in neuroscience by applying the same mathematical rules.

The exciting proposition is that this new framework offers the opportunity to allow the data from self-directed neuroplasticity experiments to be understood in a coherent, scientific and useful way. More so than previous frameworks.

Hoping you are still with me, let's explore some more of this new way of looking at the world. Rather than thinking of the world as made out of lots of bits of matter, consider instead that it is made up of potentialities.

What is quantum theory?

Niels Bohr said: 'Those who are not shocked when they first come across quantum theory cannot possibly have understood it' (Bohr, 1995). As you explore this section please bear this in mind. An important point is that contemporary physics, quantum physics, does not replace classical Newtonian physics, it adds another layer of understanding. It came about because the Newtonian principles were not working when physicists were exploring the subatomic world. The majority of physicists would say that quantum physics *only* applies to the microscopic. However, as we have seen, there are some scientists who believe that within time we will understand the macroscopic world through this lens also.

There are some occurrences at the quantum level that may challenge your current understanding of reality. If you took any physics at school it was likely to be classical physics and you were probably told that atoms – the small building blocks of matter – are mostly empty space. The analogy here is that if a basketball is the nucleus of a hydrogen atom then the electron that representationally circles around it would be 20 miles away. Everything in between is empty. Now, however, it is understood that this space is filled with powerful energy.

Another discovery that challenges widespread understandings is how electrons move from orbit to orbit. They jump. They disappear from one place in space and appear in another; a quantum jump. Even more interestingly, we cannot predict where the electron will reappear or when they will make the jump. We can only formulate the probability of these things.

Here are some key things to be aware of from the world of quantum theory:

- the double slit experiment;
- Heisenberg's uncertainty principle;
- the Einstein, Podolsky and Rosen (EPR) paradox;
- the Copenhagen interpretation.

The double slit experiment

This experiment is a fairly simple one in construct, but mind boggling as we try to grasp it. The experiment involved sending a beam of light through a thin plate that has either one or two slits in it. The important things to know are that:

- Particles can be considered to have mass, we can represent them in our mind as tiny balls. If we were to throw a tiny ball, covered in paint, at a wall we would see a tiny circular imprint on the wall.
  - Fundamental 1: when particles in the experiment pass through one slit they make a pattern on the wall behind that is in a line.
  - Fundamental 2: when particles pass through two slits they make a pattern that is in two lines.
- Waves do not have mass. Instead we think of them as oscillations or vibrations. It is useful to imagine dropping a coin into a water fountain and observing the ripples of waves emanate out from that central point.
  - Fundamental 3: when waves pass through one slit they make a pattern that is in a line (similar to the particle but not identical).

- Fundamental 4: when waves pass through two slits they make a very different pattern; multiple lines (this is called an interference pattern).

It was believed that light was made up of particles only, discrete balls of matter. The physics-shaking shock occurred when a particle was passed through one slit and it made the pattern in a line (expected) but then went through two slits and it made the multiple line patterns (like in fundamental 4, which was very unexpected). This meant that the light was behaving like a wave, which was not at all what Isaac Newton had led us to believe.

In 1961 this experiment was done for the first time with something other than light: electrons. Our understanding of electrons is that they are particles, so they should behave in the way that fundamentals 1 and 2 dictate. They didn't. In fact, even relatively big molecules would behave as if they were waves. The magnitude of weirdness is huge at this point. The physicists tried lots of things to work it out but they just kept discovering weirder things. The particle would act as if sometimes it went through just one slit, or just the other, or neither, or both! This is like a single paint ball being fired towards two gaps and splitting into two to pass through both gaps and then rejoining again afterwards. It didn't make any sense.

Not to be outdone by a particle, though, the physicists decided to put a measuring device by the slit in order to really see what was happening. You'll never guess what happened (seriously, it's that amazing): the particle started behaving like a particle again (like fundamental 2). It is said that the observer collapses the wave function simply by observing.

Heisenberg's uncertainty principle

This basically states that we cannot know both the location and what speed a particle has: by measuring one quality we affect another.

The Einstein, Podolsky and Rosen (EPR) paradox

The EPR paradox is fascinating. A little background is that Einstein was really not a fan of quantum physics. He once claimed: 'God does not play dice with the universe.' To which Niels Bohr responded: 'Stop telling God what to do!' Einstein, Podolsky and Rosen set out to design a thought experiment that would discredit quantum physics. The idea was that you arrange for two particles to be created at the same time, meaning they are entangled (or in superposition). You send one particle off to one side of the universe and the other to the opposite side. Then you do something to one of the particles to change its state. The other particle immediately also changes its state. This was the thought experiment that they thought was pretty crazy. Remember that Einstein in his theory of relativity stated that nothing could travel faster than light? Well this, they thought, would go against that and involve communication across a universe... in Einstein's words 'spooky action at a distance'.

John Bell came up with a theory (Bell's theorem), which suggested that the EPR paradox is exactly what happens. He suggested that our concept of things being in one place, localization, is incorrect. That in fact everything is non-local and particles are linked on some level. This level is beyond time and space. On a quantum level things are connected, entangled. Schrödinger said that entanglement is *the* interesting aspects of quantum physics. Our friend Henry Stapp called Bell's theorem 'the most profound discovery of science'.

The Copenhagen interpretation

Niels Bohr lived in Copenhagen and, along with his colleagues, pushed forward the idea that the observer has an inescapable influence on physical processes. This challenging idea that we are not neutral bystanders remains a powerful one. Bohr questioned Heisenberg's assertion that things existed before they were observed. Bohr believed that particles do not exist until we observe them. Reality is created on a quantum level by observing and measuring. Einstein could not accept this anti-common-sense way of thinking. We do have a responsibility to question how common sense became so.

Lynne McTaggart suggests: 'Reality is unset Jell-O. There's a big indeterminate sludge out there that's our potential life. And we, by our very act of involvement, our act of noticing, our observation, we get that Jell-O to set. So we're intrinsic to the whole process of reality. Our involvement creates that reality' (McTaggart, 2012).

Nerve terminals

Most neuroscientists work from the assumption that classical physics will answer the questions around how consciousness and brain processes relate to one another. In order for this to be the case, classical physics has to be applicable at the microscopic level. According to Heisenberg's uncertainty principle it would be impossible, according to the principle, to gather empirical evidence. The next step is to consider the questions we should ask in order to move forward. For example, this one from Stapp (2011): 'How important quantitatively are the effects of the uncertainty principle at the microscopic (ionic) level; and if they are important at the microscopic level, then why can this microscopic indeterminacy never propagate up to the macro level?'

You'll remember from Chapter 15 on neurons and synapses the classic model of these. The electrical signal travels down the axon to the nerve terminal and then triggers the opening of channels so that calcium ions can flow into the terminal. Here there are vesicles, which are home to neurotransmitters. The calcium ions diffuse to special sites where they then trigger the vesicles to release the neurotransmitters into the synaptic gap. The neurotransmitters communicate to the neuron on the other side of the gap.

Sir John Eccles won a Nobel Prize for helping us to understand how neural communication takes place. In 1986 he proposed that the probability of neurotransmitter release depends on quantum mechanical processes. This means that the brain chemicals – which we are becoming familiar with (dopamine, serotonin, adrenaline) – are released according to some quantum laws. Some scientists believe that an observer, in this case your own mind, can influence these. In grossly simplified terms, this implies that the chemicals that flood and impact our body are affected by quantum processes, which are subject to the effect of the observer. You have an impact on the probability of your mind communicating things.

This next section gets a little more technical than we have explored before, but is key to the core concepts of the quantum brain. The channels that the calcium ions use to get into the nerve terminal are called ion channels. They are small, at some points only measuring a nanometre! This is on a similar scale to the slits in the double slit experiment. There is quantum uncertainty that any one calcium ion will be absorbed on the triggering site. Subsequently there is empirical uncertainty of a neurotransmitter being released that is affected by the quantum mixture of states of uncertainty of all the calcium ions.

It is suggested that this quantum opportunity (Stapp calls it quantum splitting) propagates to neuronal behaviour then to the behaviour of the whole brain. The argument being that bifurcation points arise where some quantum potentialities go one way and some go another. Stapp (2011) says: 'any claim that the large effects of the uncertainty principle at the synaptic level can never lead to quantum mixtures of macroscopically different states cannot be rationally justified'.

The idea that intention and attention have an impact in the brain are familiar in the fields of psychology, religion, philosophy and others but relatively new to neuroscience. The consideration that 'effort of attention' has an impact on brain activity and therefore our behaviour is a powerful one. Intentional inputs are important. Harold Pashler, a distinguished professor of psychology, mentions experiments and research that feeds into these discussions from a psychological perspective (Pashler, 1999). When a person is tasked with doing an IQ test and also giving a foot response to rapidly presented tones their mental age goes from adult to 8 years old. We find time and time again that reducing effort of attention has consequences. Mental exertion reduces the amount of physical force that a person can apply. Pashler doesn't explain these occurrences in terms of quantum physics, but we can see the links.

In other experiments, this time led by Ochsner, participants are also being put to work (Ochsner, 2002). Here the participants are trained to cognitively re-evaluate emotional scenes that they are shown by consciously creating and holding on to an alternative fictional story of what is happening. They were initially shown either a negative or neutral photo for four seconds. During this time they generated a normal response to it. Then they were instructed to 'reappraise' or 'attend' to the negative or neutral photo also for four seconds. Afterwards they were asked whether they had reinterpreted the photo in order to reappraise it, or if they had used another strategy.

The accompanying fMRI scanning results showed that reappraisal is correlated with increased activity in the left lateral prefrontal cortex and the dorsal medial prefrontal cortex and decreased activity in the amygdala and medial orbitofrontal cortex. Stapp reports that, according to classical physics, when we do something like the above experiment and exert conscious effort the feeling that is linked to the process is an epiphenomenal side-effect. This has no effect on the brain. The idea that it has an effect on one's thinking is not in alignment with classical physics. The challenge is that at the micro level we know that classical physics in the brain does not fully work. In Stapp's words: 'The presumption that it gets restored at the macro level is wishful and unprovable' (Stapp, 2011)

Where 'the quantum brain' will be in 5, 10 and 50 years' time is largely up to the current researchers who are able to stand on the shoulders of the researchers before them to take us forward. I believe it is likely that we will continue to see more of science that demonstrates the power of our mind to affect our brain. In the meantime there are many powerful studies that add depth to what coaches are currently doing.

PART SIX
Neuroscience of classic coaching areas

Part Six focuses on what neuroscience can share with us about classic coaching tools, techniques and concepts. These are mostly things that a coach will be familiar with and utilize in their work regularly. It can be helpful to remember that neuroscience is a field in its own right, not set up or even focused on finding things out just to back up what coaches already think and do. Therefore, sometimes we need to be careful not to be shoehorning the science into a construct that it just doesn't fit into. Keeping an open mind is a good way to move forward most effectively.

Topics such as goals and beliefs are likely to underpin work with almost everyone regardless of the style or methods that a coach uses. A manager or leader in an organization responsible for coaching individuals will at least have conversations around goals, and what is thought about abilities and possibilities.

Then we have topics that as a coach you almost certainly are aware of and touch on through the coaching process. Here we look at what is going on in the brain with these things and what that means. Often we see just how important a more detailed understanding of them is. For example, with self-control and willpower we know that the achievement of many goals and the formation of new habits often relies on good self-control, at least initially. In order to help best those we work with, and set them up for success, looking at what the research is telling us seems obvious.

Through our Neuroscience for Coaches programme we have seen that not all coaches are familiar with concepts such as *flow* or *mindfulness*. Mindfulness is accruing a large body of neuroscientific research to support the understanding of how it has such positive benefits for individuals. Our trained coaches tell us that it gives them a sense of credibility when they can explain the research that backs up what they are doing with their clients.

Self-control/ willpower

What is it?

Self-control or the utilization of willpower involves the ability to consciously decide what you will do and when, to resist temptation, and to put off a reward until later.

Why is it important for coaches?

Your clients' ability to control themselves is key to their success. The construct of many goals involves the exercising of self-control. Sometimes just getting started on a goal requires willpower. For example, for someone who wants to run a marathon but is not that keen on running, they may need to exercise self-control to make themselves go for a run on their planned days rather than go out for a drink with colleagues. Once they have been on a few runs and experienced the flood of endorphins, and felt great as a result, they may find themselves keener to go out for a run than to do other things. Less self-control will be required.

There are other goals that may require self-control throughout. For example, attempting to address a habit where someone regularly interrupts colleagues can be a constant challenge. It will take willpower and self-control to mentally and physically hit pause before the interruption has occurred and the damage to the relationship has been done. Understanding the latest findings on self-control can help you to set up your client with the best chance of succeeding.

A lack of self-control can predispose the client to create excuses for lack of achievement. If a person repeatedly does not achieve something, then rather than take responsibility for the reality of the situation it can be natural for some people to start making excuses. In the long term this can damage a person's chances of achieving future goals. Addressing any challenges around self-control in a safe and calm environment is a great investment for anyone you are working with.

What is the underpinning neuroscience?

Research has shown that exercising and depleting willpower causes a slow-down in the anterior cingulate cortex. You might remember from Chapter 6, where we looked at the anterior cingulate cortex, that this area is vital to self-control. It is commonly known for its conflict-monitoring activities. It is the area that picks up any mismatches between what you intend to do and what you actually do. (All those times you've put the toothpaste in the fridge while multitasking making your morning coffee... your ACC has picked up on it! It knows when you are making an error in the moment.)

The disappointing news is that a review of the scientific literature showed that there were strong behavioural indicators when willpower was depleted, but weak feeling indicators. This makes it slightly trickier for your client to know when they are about to snap at a colleague, for example, or lapse back into smoking.

The prefrontal cortex is also very important in controlling ourselves. The PFC has vital roles in planning, our working memory, organization, anticipation of consequences and controlling impulses. Often considered at the other end of the spectrum is the amygdala, which has a great deal of power over our decision making. Sometimes there is conflict between the rational cognitive parts of the brain and the emotive reactive parts.

What are any interesting studies in this area?

One of the pivotal studies in the field of delayed gratification is the famous Stanford Marshmallow Experiment (Mischel *et al*, 1970). This involved a series of children being sat in a room with a marshmallow on the desk in front of them. They were told that they could eat the marshmallow at any time or wait for the experimenter to return and then they would receive an additional marshmallow to enjoy. They found that most of the children couldn't delay their gratification long enough, although watching them try is both illuminating and comical.

Where the implications develop are when Mischel followed up many of the initial subjects many years later. His daughters, who were in the same school as lots of the children, would often tell him stories about how their classmates were doing. This prompted him to investigate further. He found that the children who were able to resist the marshmallow longest were achieving better grades, were more popular, eventually earned higher salaries and had a lower body mass index.

Brain differences have been found between those who can resist the marshmallow and those who give in to temptation. The key differences were found in the prefrontal cortex and ventral striatum. The study looked at 59 of the adults who had been in the original marshmallow experiment as children and represented people from both extremes of the ability to delay gratification (so the really good people and the really poor people). The nature of the test was changed to a more adult version, so participants were asked to signal when they saw a happy or frightened face. Dr Casey from the study explained: 'In this test, a happy face took the place of the marshmallow. The positive social cue interfered with the low delayer's ability to suppress his or her actions' (Casey *et al*, 2011). So when individuals with lower abilities to delay gratification saw the happy face they pressed the button when they shouldn't. The emotional connection of a happy face interfered with the rational instruction of when to push the button.

When the participants performed this test in the fMRI scanner the results showed that in the people who were able to delay gratification for longer the prefrontal cortex was more active. In those who struggled to delay gratification it was the ventral striatum that was more active.

Roy Baumeister has performed many interesting studies. One involved fasting college students. These hungry people were first left in a room with cookies, chocolates and radishes. Some were invited to eat the yummy things. Next they were given geometry puzzles to try to complete, although they were insolvable. (How long it takes them to give up is a standard technique to measure perseverance.)

The students who had been allowed to indulge in the cookies and chocolates kept working at the puzzles for around 20 minutes. This was the same as a control group (hungry but not tempted). The group only allowed to eat the radishes gave up after eight minutes. This tipped off the researchers to considering that willpower got depleted. This has big implications for how we optimally organize ourselves.

Another experiment required some subjects to suppress their emotions and others to amplify them while watching a sad film. The control group just watched it normally. Their stamina afterwards was tested (using the standard hand exerciser squeeze test) and unsurprisingly it showed that suppressing and amplifying emotions resulted in lower stamina.

In other studies Baumeister and colleagues (1998) suggest that blood sugar levels may drive the strength of willpower. This controversial idea states that the lower your blood glucose level the more likely your willpower is to fail you. Several of the studies in this series involved individuals doing lots of effortful cognitive tasks. Their willpower is then tested in some way, and they tend to fail. Those who are given a sugary lemon drink tend to have stronger willpower than those who get the sugar substitute in their lemon drink. This may link to the study above, where the students who ate cookies spent longer on the puzzles.

Carol Dweck *et al* (2013) have proposed that rather than the key being sugar, it is in fact our mindset that has a key impact on our willpower. They suggest that if we believe sugar will help then it probably will. Researchers at Stanford University explored people's beliefs about willpower and tested whether and when these individuals experienced a dip. They found that those who believed willpower was abundant didn't seem to need sugar in order to persevere through challenging tasks. Baumeister has also shown that people in good moods have stronger willpower.

We can often learn things about willpower from people who are unwell. It would be interesting to explore, for example, how people with anorexia nervosa appear to exhibit such strong willpower around resisting something as key to humans as food. Their sugar levels are low, and yet something enables them to retain a high level of self-control. By understanding this mechanism better it could provide additional help to people experiencing weak or exceptionally strong willpower when either of these is unhelpful to the individual.

These insights bring us back to a place of hope. We have hope that neuroscience will drill down into a more detailed understanding of how willpower works so that we have a more thorough understanding of how to improve it. Trial and error, as is often the case in such psychological studies as those we have covered here, can only take us so far. At times we are seeing seemingly conflicting information but the drive for a deeper understanding is spurring on every discipline involved in exploring this field.

What can I do with a client this afternoon?

Work with your client on ways to strengthen their mind's 'breaking' abilities and help them to come up with some creative ways to distract themselves. An example of this would be if someone was working on getting a promotion and had been told that they needed to speak out more during meetings. Perhaps the natural tendency of your client is to sit quietly in meetings. Internally they are having a great dialogue... but no one else gets to hear it. What might come into their mind just before they are about to speak is thoughts of what others may respond with, or questions that people may ask. This may result in an ongoing internal dialogue until their point no longer fits the external conversation that has continued. Maybe your client could experiment with having the numbers one to five written on a piece of paper in front of them during meetings. When a comment comes into their head they make a one-word note next to number one and then speak their comment out loud. This could act as a physical distraction and mind-breaking opportunity. Their aim during each meeting could be to get to five comments.

Suggest that your client commits to doing one thing at a time that requires willpower, ideally when they have not got lots of other cognitively draining things also going on. So if they were considering embarking on a new and challenging goal then check in on what else is going on in life at this time too. If you are a manager and someone wants to start a new project that will involve them working with a colleague where there has been conflict in the past, then careful self-control management would increase the chances of success. If this person is working on several challenging things at once then they may find it harder to exercise self-control.

Perhaps task your client with making a to-do list in order to get things out of their brain. This could reduce the cognitive load they are carrying around and might enable them to exercise more willpower when they need it.

Habits

What is it?

Habits are regular tendencies or practices. They are often unconscious and automatic. We each have lots of them, some useful, some not so.

Why is it important for coaches?

Habits are naturally great things that free your client's mind to focus on new things, making them both more efficient and productive. They enable us to cope with the huge amount of information, choices, decisions, goals and emotions we experience every day. We have habits for a large part of our daily life. Many habits are unconscious, which means they can be directing your client's behaviour without them being consciously aware of this. While good habits make it easier to achieve desired results, because we naturally do these things without conscious thought and less energy is required to fall on these defaults, undesirable habits make it difficult to achieve results.

What is the underpinning neuroscience?

One of the important things to be aware of here is called Hebbian theory, which underpins Hebbian learning. A Canadian neuropsychologist called Donald Hebb theorized that:

> When an axon of cell A is near enough to excite cell B and repeatedly or persistently takes part in firing it, some growth process or metabolic change takes place in one or both cells such that A's efficiency, as one of the cells firing B, is increased.
>
> (Hebb, 2002)

This has been simplified to 'Neurons that fire together, wire together.' As you have repetitive thoughts or take actions repeatedly your neurons fire again and again, forming a stronger and stronger neuronal circuit.

The formation of these circuits has the benefit of meaning we can do things on autopilot. The downside is that many of the circuits that are created are undesirable to your clients. The more that a habit is performed the stronger it becomes. This is because when the neurons fire they attract a protein blend called myelin to insulate it. This speeds up the transmission of signals, in effect making it easier and quicker for the habit to occur next time.

The neural pathways that encode habits are found in the basal ganglia, so this is the part of the brain often associated with habits. However, recently it was discovered that habits are not exclusively automatic. A study from MIT showed that a small part of the prefrontal cortex, called the infralimbic cortex, is involved in shutting off habitual behaviour (Smith *et al*, 2012).

What are any interesting studies in this area?

One area of research linked to habits is that of people with addictions. It has been claimed that people with gambling addictions, for example, have habits that they have no control over. Reza Habib, a cognitive neuroscientist, did some research that increased our understanding in this area (Habib *et al*, 2010). The experiment involved some serious gamblers whose lives were being affected by this habit. These people had lied to their families and had missed work in order to gamble. The other people in the study gambled so-cially but did not have any problems associated with this habit. Each person went into an fMRI scanner and watched the wheels of a gambling machine turn around on a video screen. This slot machine would either deliver a win, a loss or a near miss. There was no money involved in this experiment; the idea was that the gamblers simply watched the screen while their brain activity was recorded.

What they found was that the brains of the pathological gamblers got more excited about winning. The people in the experiment knew they weren't actually winning any money in this instance, but their brains still responded similarly. The areas related to emotion and rewards were more activated in the pathological gamblers than in the social gamblers. What the researchers found even more interesting was what happened in the brains of the two types of gamblers when there was a near miss on the screen. The pathological gamblers' brains registered the near misses very closely to how they registered a win. The social gambler registered the near miss similarly as a loss.

As we know as coaches, different people can view a single event very differently. There is corresponding neurological activity that underpins these differences. What may on the surface seem simple can be more complex when we cross-reference with what the brain is doing. The responses of our brains can trigger habits. So when a gambler wins and gets a mental high then they are triggered to play again to get another high. For pathological gamblers, the same is unfortunately true when they get a near win.

A key thing for consideration when working with habits is the preparation for what happens when an old habit is reverted to. If people don't know how habits work then there is a tendency to feel that they have failed, to feel worried they'll never be able to stick to the new habit or simply to feel guilty if an old habit creeps in.

Ivan Pavlov did some interesting research that gives us an insight into why habits sometimes return. In one set of experiments he found that if a rat was conditioned with a tone and electric shock in one box, and then moved into another box, the fear response could be 'removed'. If the rat was put back into the original box then when it hears the tone again it would experience the fear response again. It was also found that stress could reinstate a previously extinguished response.

Basically, habits tend to be deeply ingrained. When you stop performing a habit, such as snapping at a colleague whenever they interrupt you, then this neural circuit gets weaker. However, we don't believe it disappears completely. When you create a new habit and strengthen it over time and with lots of repetition, for example responding to any interruptions with a hand indication and the request for the person to 'give me a minute', then under normal circumstances this becomes the natural response. However, when under stress it is possible that the old circuit becomes favoured again and a return to old behaviour is experienced.

Research from the National Institute on Alcohol Abuse and Alcoholism offers some additional insights into breaking habits (Crews *et al*, 2009). This ability is hugely valuable to people both at work and everywhere else in life. By being able to create and enact new habits rather than default to old ones individuals can access flexible decision making and can align more effectively with goals. You might remember that other research has shown that the dorsal medial striatum is involved in goal-directed actions and the dorsal lateral striatum is needed in habitual actions. This particular study found that the orbital frontal cortex is critical in shifting from habitual behaviours to goal-directed actions. This is an area of the brain that is involved in obsessive-compulsive disorder, where unhelpful habitual actions are adhered to compulsively. This study was performed using mice, and the process called optogenetics, enabling scientists to activate individual neurons.

An experiment conducted at MIT looked at this area of breaking habits, this time involving rats. The study first conditioned rats to run a maze in a specific way. They checked that the habit of running the maze in this specific way had been thoroughly ingrained before moving on to the next stage. Normally we think of habits being linked to the basal ganglia, which they are. There is also involvement of a small area of the prefrontal cortex known as the infralimbic cortex. Using this advanced technique again, optogenetics, the researchers were able to inhibit specific neurons using light in the infralimbic cortex for seconds at a point where the rats were deciding which path to take next. The effect was that the rats stopped their habit and chose to run a different way. One researcher said that by turning off the infralimbic cortex it switches the rats' brains from an 'automatic, reflexive mode to a mode that's more cognitive or engaged in the goal'. This is good news for coaches because it means that there is a degree of executive function involved in breaking habits.

When the rats formed a new habit the researchers found they could switch this one off too by again inhibiting the infralimbic cortex. The rats reverted to the previous habit. This is evidence that old habits are not forgotten, there is a representation of the habit still intact within the brain and this can be retrieved instantaneously.

What can I do with a client this afternoon?

Due to the nature of habits there are lots of things that you can work on with your clients. There are also lots of links to other concepts within coaching. A great approach, depending on your client and the situation, could be to focus on a specific habit. In order to decide where to focus you could:

1 Invite your client to verbally share with you five habits that they feel are holding them back or doing them a disservice in some way.

2 Once you feel that you understand these habits, the triggers for them and the negative effect they are having, you could then move on to get clear on what habits would be supportive in reaching desired goals. Perhaps elicit between 3 and 10 here. You could ask your client to write these down either during a session or, even better, at home. Again, you want to know the value that these habits will bring to the individual. The next stage is like being an architect: the aim is to identify all the small parts to one of the new desirable habits.

3 Identify small steps involved in creating a new habit.

Then, depending on how successful your client normally is with establishing new habits, you want them to start implementing and mastering one step at a time. So this stage can be really slowed down for people who might have a history of not creating solid new habits – literally taking one component and working on it until it is second nature before introducing another step. Another point to mention here is that this process is best initiated when life is not overly stressful. If your client is trying to move house, lose weight and start coaching their team at work in a new way then it would be wise to wait until the house move is complete and the weight lost.

There are a few other things you could do with someone you are working with that could be useful:

- Help them understand how a habit is formed and therefore how to avoid giving attention to undesirable habits.

- Give comparative frames of reference of people who have rewired their brains: for example, people who have regained the use of paralysed limbs (this helps expand what your client thinks is possible). Check out **www.neuroscienceforcoaches.com** for more ideas and resources to use with clients.

- Explain the neuroplastic nature of the brain so that they are clear that habits can be changed.

Optimism

What is it?

Optimism is about having a positive attitude towards the future. There are two main types of optimism that are of interest to us. The first is blind optimism. This is the classic 'everything will be great' response, even when you have been gaining weight for the last two months and your clothes no longer fit – a scenario where this type of optimism hasn't been shown to be overly helpful. The second type is called realistic optimism. This is more the attitude of: 'If I make some changes then things will be great, making those changes may be difficult, but in the long run it will be worth it.' Ultimately, people who have a realistically optimistic approach to life tend to believe that they will be able to cope with the challenging or bad things that will happen.

Why it is important for coaches?

- The field of personal development *can* be very pro rose-tinted glasses and so it is important to understand what this means.
- How we interpret life events can affect our life's trajectory.
- Equipping clients to be mentally robust and successful.
- Being subject to the optimism bias can have positive or negative effects – and being aware of these increases choice.

What is the underpinning neuroscience?

It is fairly widely accepted that generally we see different levels of activity in the left and right brain hemispheres (this is called cerebral asymmetry) when people are in positive versus negative states. This can be seen when we use electroencephalography (EEG) with people. It has also been shown that subliminal biases in attention and interpretation can be linked to an increased vulnerability to stress. We know that the left inferior frontal gyrus responds to good news and the right inferior frontal gyrus responds to bad news.

It has been shown through fMRI scans that when we think about upcoming positive future events this activates both our rostral anterior cingulate cortex and the right amygdala. It is possible that we are helped to imagine these future events by assessing emotions from past events. It is also thought that these two brain areas work together to downplay negative emotional responses. This could help us to be optimistic and driven to achieve our next set of goals.

What are any interesting studies in this area?

There are many fascinating studies on this topic. What follows are a couple of these studies along with an insight into the 'optimism bias'. Suzanne Segerstrom (2007) has done a particular experiment that looks at the behaviour of optimistic people and how it differs from pessimists. A group of people are divided into the optimists and the pessimists. They are then all given a variety of different anagrams, including an impossible one. The findings show that the optimists spend more time trying to solve it. It is hypothesized that this trait of being persistent stands the optimists in good stead for life, enabling them to achieve more – simply through being committed to doing more. Obviously persistence is a good trait only under certain circumstances. For example, for the entrepreneur who keeps knocking on doors, such as Colonel Sanders or J K Rowling, persistence is what enabled them to finally succeed. However, for firefighters who thought if they persisted for just a few more seconds, despite their instincts or their facts telling them otherwise, the outcome has proved to be very different.

Colin MacLeod *et al* (2002) from the University of Western Australia performed a very insightful study that could prove very useful to coaches who want to help people change subliminal habits. MacLeod's experiment tried to train students to either notice or ignore threatening words. They were shown both a nasty word and a neutral word at the same time on a screen. A split second later they were shown a target letter and they had to respond as quickly as possible by pressing a button. The experiment was set up so that for half of the students the letter always appeared in the space of the threatening word and for the other half in the place of the neutral word.

These students were then given anagrams to solve, some of which were impossible (a trusted way of stressing people!). Those who had been trained to notice threat reported feeling more stressed during this process. This is an indicator that focusing on negative rather than neutral things could affect our daily activities. If any of your clients need to reduce their stress levels then try looking at their anchors. You may have encountered anchors as a cognitive bias that predisposes us to rely heavily on existing information.

Like the ship's device to keep us secure to one place, an anchor can tie us to thoughts, actions or even decisions too. Things may be masquerading as pessimism that is instead learned responses.

A practical demonstration of a positive application of this work involved a group of students who were relocating. This could have extended applications to any organizations that routinely relocate employees. In this group of students, half of them received training to help them avoid threat and this half showed less stress and anxiety when they moved from Singapore to Australia. The classic sentiment of 'we get what we focus on' comes to mind here.

The optimism bias

Tali Sharot (2012) has researched extensively what is referred to as 'the optimism bias'. In essence, this is the tendency that most healthy individuals display to presume that bad things won't happen to them and that good things will. So people tend to underestimate the likelihood that they will get divorced, suffer from cancer or be involved in a car accident. It also affects our ability to predict things such as how long a project will take (ever find yourself taking longer than you thought over things?) and even how much we will enjoy a holiday.

People used to think that having low expectations was the key to happiness. This hasn't been shown to be the case. In fact a study by George Lowenstein (1987) showed that anticipation is powerful. He asked a group of students how much they would pay to receive a kiss from a celebrity over different time spans (so, for a kiss right now, in one day, three days, one year, etc). The results were surprising: the maximum amounts fetched were actually for in three days' time, not right now. So the pleasure of anticipation was valuable to people. Interestingly, people seem to prefer Fridays to Sundays (a day of work rather than a day of rest or play). Could this be linked to the absence of the good anticipation of tomorrow, the dreaded Monday?

Optimism has been shown to make you try harder. Interestingly, when people are given new information, such as the statistics of how many marriages end in divorce, not everyone updates their beliefs to reflect this data. Research has found that people adjust more in response to positive information, so if a person thought they had a 60 per cent chance of getting divorced, and then found out it was only a 40 per cent chance, they would make a bigger adjustment than if they believed they were only 10 per cent likely to divorce and found out the statistics pointed towards a 40 per cent chance. At these times, the frontal lobes are not effectively coding errors in predictions. Optimism appears to reduce our ability to update ourselves based on undesirable information regarding the future.

Is optimism good for us?

People who don't have positive expectations about the future are more likely to have symptoms of mild depression and anxiety. Subsequently, some people say that optimism is vital to our mental health. Some studies also point to the benefits to our physical health, which include living longer and more healthily. Even in those people who do get sick, such as with cancer or AIDS, the survival time of these people who are also optimists is greater than the pessimists.

Optimists enjoy reduced stress levels (which causes an overactivation of the autonomic nervous system and hypothalamic-pituitary-adrenocortical axis). They catch fewer infections and have stronger immune systems. It has also been suggested that this positive outlook promotes healthier behaviour, so exercise and healthier food choices. Even in the professional world optimists come out on top with higher pay than others.

Downsides

Optimism can pose risks, including people behaving in ways that are not as precautionary as would be good for them. For example, having unsafe sex, not attending medical screenings, not having vaccinations or being adequately insured. Extreme optimists are more likely than mild optimists to smoke and fail to put aside savings.

What can I do with a client this afternoon?

- Explain what optimism is and what the benefits and risks with it are.
- Support them to brainstorm anywhere that they are unhelpfully exhibiting the optimism bias.
- Help your client to connect to the pleasures of anticipation. Suggest they make a list of everything that is happening this month that they are looking forward to. Perhaps each night they can choose one of these things to do a couple of minutes of mental simulation around, imagining what it will be like.

If your clients identify themselves as pessimists and express a wish to change, then things that may help are:

- Mindfulness-based therapy – these courses are meditation-based and normally run for around eight weeks. Several centres are now offering them in the United Kingdom and in other parts of the world they are even more prominent.

- Cognitive-bias modification – there are many ways to do this now available in the marketplace.
- The practice of 'an attitude of gratitude' where you suggest your client does a variation of finishing the day with three things they are grateful for – an exercise that has roots in this way of thinking.

Goals

What are they?

A goal is a desired result. Goals can be traditionally tangible or intangible. There are ways to measure both and they can both be brain friendly. Some researchers propose that the attainment of goals relies on three core processes: goal maintenance, performance monitoring and response inhibition. Goal maintenance is one's ability to keep a cognitive representation of that goal in mind. Performance monitoring is the ability to keep tabs on the end state and current state, and identify any lapses in progress. Response inhibition is about being able to stop those behaviours that aren't in alignment with moving towards the goal.

Why are they important for coaches?

- Goals are at the heart of coaching.
- Not achieving a goal can affect a client's mindset negatively.
- Repeatedly achieving goals can make it easier to continue doing so.

What is the underpinning neuroscience?

The neuroscience of goals is a huge field. Our brains are clearly important in the setting, pursuing and achieving of goals. The scope of what is involved in achieving a goal is very wide. Achieving a goal through taking appropriate action has components associated with it:

- deciding on the goal;
- thinking creatively;
- planning how you will achieve it;
- visualizing the goal, and the steps to reach it;
- paying attention to/being affected by relevant cues – conscious and unconscious programming (anchoring, nudging, priming);

- exercising willpower or self-control;
- helpful habits;
- monitoring progress and adapting;
- inhibiting internal distractions (emotions);
- inhibiting external distractions;
- taking action.

Having the backing of neural networks set up to best support your clients in achieving their goals is really important. These can be set up and strengthened in a variety of ways. Not paying attention to this core level of programming is setting people up to have more challenges than they need to.

The existing research points us in the direction of certain brain areas as being involved with goal-related processes in general. These are the dorso-lateral prefrontal cortex (dlPFC), ventrolateral prefrontal cortex (vlPFC), dorsal anterior cingulate cortex (dACC) and parts of the striatum. When we get to the point of actually taking action it is the regions responsible for motor control that get involved. Namely these include the primary motor cortex, supplementary motor area, premotor cortex, cerebellum and basal ganglia. The basal ganglia seems to act as a motor control hub in integrating information from inputs and then outputting in response.

What are any interesting studies in this area?

There aren't many neurocognitive studies that address goal pursuit in its entirety, but several studies look at different components of it. It is quite difficult from a study design perspective to take into account the complexity of goal pursuit and to study this with tools like functional magnetic resonance imaging (fMRI). Here we consider several studies that look at individual components of working towards goals and then one that looks at the coming together of goal maintenance, performance monitoring and response inhibition.

Overall, many studies lead us to believe that the prefrontal cortex is involved in creating and maintaining goal representations. We can also reliably say that different parts of the PFC do different things within this area. Goals can be both extrinsic (typically, given to us) and intrinsic. The difference is paralleled by a neural distinction of lateral and medial prefrontal cortex networks, as seen in Lieberman's work (Lieberman *et al*, 2007). In 2006 Johnson led a study that showed greater medial prefrontal cortical activation while the subjects reflected upon their personal hopes and aspirations rather than distraction (Johnson *et al*, 2006). This has links to internal goals conceptually. On the other hand, extrinsic goals are linked to the lateral PFC.

The classic Libet experiment, which showed that a readiness potential in the brain occurred 500–1,000 milliseconds prior to the conscious experience of intending movement – has links to goal-realizing behaviour. This is almost like our brain knowing that we intend to do something before we do. It has raised a lot of dialogue around the notion of free will. Inhibiting responses is vital to the achievement of goals. For example, stopping ourselves sitting down on the sofa before going for that run, or stopping ourselves snapping at that colleague before we have considered a calm response. Conscious inhibition relies on the prefrontal cortex.

A study by Spunt and Lieberman (2014) highlighted some important things to be aware of practically with goal setting. They recorded the brain activity of subjects who were watching videos of people doing things such as brushing their teeth or reading a newspaper. The subjects were told to think about how the people were doing the tasks and then why they were doing them. The 'how' thoughts corresponded with areas on the left side of the brain involved in planning movement and tracking location (the premotor cortex and posterior parietal cortex). When they considered 'why' then areas involved in thinking about states and intentions of others lit up (the right temporoparietal junction, the precuneus and dorsomedial prefrontal cortex). Considering different elements of a goal activates different areas of the brain. We look at what you as a coach should do about this in the next section of this chapter.

The study 'Interactive effects of three core goal pursuit processes on brain control systems' by Berkman, Falk and Lieberman (2012) identified functional activations distinct to each of the processes. They also reported interactive effects between response inhibition and goal maintenance in the dorsal anterior cingulate cortex and between performance monitoring and goal maintenance in the superior frontal gyrus and supramarginal gyrus.

When we focus in on our ability to maintain a representation of the goal in our working memory so that we can take action linked to that goal, we are talking about goal maintenance. The Stroop task, which normally activates the dlPFC and ACC, gives us some more interesting information. The Stroop task involves people naming either the word they see or the colour of the word. This test is actually requiring participants to do both goal maintenance and response inhibition (when the word and colour of the word cease to be the same). For example, the word red will be shown in red, the word yellow in yellow, then perhaps the word red will be shown but in blue. What we see is that the dlPFC is only active when doing the first part of the test, where goal maintenance is occurring (not response inhibition).

When we focus in on performance monitoring we know that we are asking the brain to identify gaps between where we are and where we want to be. It requires us to adapt and proactively get back on course. The dACC was shown to be involved in both goal discrepancy detection and reduction. More recently the dACC has been shown to be more active when it responds to an incongruence in a sea of congruencies (eg the word red appearing in the colour blue). Now it is understood that the ACC activation signals the discrepancy but tells other brain areas to reduce that discrepancy – almost like a head bouncer who sees conflict brewing and then gets on his walkie-talkie to tell other bouncers to go in and sort it out.

When focusing on response inhibition, evidence points us in the direction of the right inferior frontal gyrus (rIFG). With studies that involve inhibitory tasks (eg task switching, go/no-go stop signal) it is thought that the rIFG is the only brain area necessary for inhibition. However, several other areas are also activated, including the dlPFC, vPFC and ACC. Newer findings suggest that the anterior insula and parts of the striatum (head of caudate) are involved in inhibition.

Implementations intention

Frank Wieber and Peter Gollwitzer propose 'implementation intentions as an easily applicable planning strategy that can help overcome procrastination by automating action control' (Gollwitzer and Wieber, 2010). In other words, implementation intentions are a valuable tool to support clients who are working towards goals. They are important in specifying the details that will underpin the achievement of a goal. Let's take a seemingly simple goal of flossing your teeth. Many people share the goal of flossing their teeth regularly; however, far fewer do it as often as they would like. An implementation intention can help, especially one in an 'if-then' format. Consider the benefits of programming yourself to pick up the floss every time you put toothpaste on your toothbrush. So 'if I put toothpaste on my toothbrush, then I put it down and pick up my floss and use it'. What about when you feel in a hurry and are tempted to skip flossing to save time: 'If I'm in a hurry then I floss while thinking about the time I'll save from not getting tooth ache and having long-drawn-out dental treatments.' The great thing about 'if-then' intentions are that you can create them for many eventualities, and indeed considering these eventualities will stand you in better stead for dealing with them.

Implementation intentions come with a wide body of research that shows their benefits. Studies have shown that they have effects over and above motivation to succeed. This was demonstrated when participants who formed temptation-inhibiting implementation intentions outperformed the groups who didn't have these in place. Other studies have shown that they can help overcome self-regulation depletion (or willpower burnout). Participants of one study had to suppress their laughter during a funny film; afterwards they were supposed to do a task involving self-regulation (solving a series of anagrams). Normally you find that after people have been exercising their willpower they get depleted and find such tasks to be more difficult, so they would solve fewer anagrams. When the participants are given an 'if-then' implementation intention, such as 'if I solve an anagram, then I will immediately start to work on the next one', then they were able to solve as many anagrams as the group who hadn't been depleted.

Process versus outcome mental simulations

Most coaches are experientially familiar with the impact of thoughts on action. One study looked at the difference between mental simulations that focused on the process required to achieve a specific goal versus the outcome of the goal itself. This study was done with college freshers for five to seven days prior to an exam. They were split into three groups and mentally simulated, by either having good study habits or getting a good grade or both. Results showed that by focusing on the process the students got better grades; the way they did that – it is suggested – was by enhanced planning, studying and reduced anxiety.

Older studies have shown that by imagining events (rather than just reading about them, for example) people have greater confidence that they will occur. We know that the brain and body respond when we imagine things. Having an emotional response to the mental simulation of the process required to achieve a goal, or the goal itself, seems to be a positive thing. It can serve to motivate clients, give valuable feedback and lead to improved action.

Willpower

The neuroscience of goals would not be complete without looking at the topic of willpower. We cover this in depth in Chapter 24, but mention it again here because it is so important to the current discussion on goals. Here we will just reinforce that a person's ability to take the small steps that will lead to the achievement of a goal can be the difference between success and failure. Perhaps an initial goal for any coaching client should be to improve their self-control! A study that looked at 92 adults and their capacity for self-control showed that it is possible to improve this by regularly practising small acts of self-control. This study involved the participants cutting back on sweets, or squeezing a handgrip, things that required their self-control. Their self-control was assessed before and after a two-week period using the stop signal paradigm. Something as simple as switching to clean your teeth with your other hand can help improve your self-control.

What can I do with a client this afternoon?

- Setting goals that are widely unrealistic can motivate some people but unmet goals can be very demotivating to other people. So, initially, the safe bet is to support clients in setting realistic goals so that they get some wins under their belt. Offer clients the opportunity to increase their self-control by working on it for a two-week period, explaining that this should make it easier for them to achieve their goals.

- Within a full goal you need to know why you want to achieve it, what you are seeking to achieve and how it will be achieved. These should be separate components, as the brain struggles to focus on them all at the same time.

- It can be useful to move up and down within the goal hierarchy from 'why' to 'how', in order to get unstuck from a task or to become more practical by focusing on the 'how'.

- Another thing we can take from the research into goal maintenance, performance monitoring and response inhibition is that it could be useful to keep goals in mind externally rather than relying on working memory. So a smoker who wants to quit might find that rather than having to remember this when their cravings strike, instead to have a visual cue somewhere, such as a screensaver on a phone or computer. Perhaps a picture or comment placed actually on a cigarette box may help. Similarly for weight-loss goals, consider having a picture of either the ideal figure or part of the process (such as eating healthy snacks) on the fridge door.

For a goal like 'go to the gym regularly' there are several things a person can do to make it easier to go. One could:

- Go to the gym and do everything you can to enjoy it (play your favourite music and tell yourself how great this is making you feel).

- Connect to the sensations of stronger muscles while doing weight repetitions, feeling your legs moving quicker on the treadmill, the tightness of your abdominal muscles after your crunches (this connects you to the benefits each and every time you go, rather than just in three months' time).

- Practise thinking about the gym while getting yourself into an excited and interested state.

- Anchor the state and the thought together by listening to an upbeat song or spraying a special gym deodorant (that you keep only for using just before you drive to the gym).

This topic has links to:

- optimism;
- willpower;
- dopamine;
- neuroplasticity;
- priming;
- nudging;
- habits.

Mindfulness

What is it?

Mindfulness meditation can be thought of as the practice of attending to present moment experiences and allowing any emotions and thoughts to pass without judgement. It can involve activities such as focusing on specific physical sensations – breathing, for example, which one observes as thoughts, emotions and bodily sensations rise up and then pass away. A newer, two-component model of mindfulness has also been put forward. This involves both the regulation of attention on immediate experience and also approaching experiences (regardless of what they are) with a mindset of curiosity, openness and acceptance.

Mindfulness has some historical links to spiritual practices. Buddhists practise mindfulness and it has been linked to achieving enduring happiness. There are many good books available specifically on mindfulness so here we will focus on the neuroscience component.

Why is it important for coaches?

Being able to be present and 'in the moment' is seen as a desirable trait. In the past we might have thought that some people were just naturally more able to be that way, luckily having a more 'composed' or 'calm' personality. Now we know that it is possible to change our brain connectivity and rewire ourselves to support present-moment awareness. Mindfulness has been shown to elevate positive emotions, to have stress-reducing effects and pain-management effects.

Mindfulness meditation works in several ways:

- attention regulation;
- body awareness;
- emotion regulation;
- change in perspective on oneself.

Coaches may find that they are working with clients on goals or outcomes that a mindfulness meditation habit would support the client in achieving.

Overall it appears that anyone wanting to use their brain for cognitively challenging tasks (so anyone working in a knowledge-based role) could benefit from mindfulness.

Mindfulness is well established as a psychotherapeutic intervention for anxiety, depression, substance abuse, eating disorders and chronic pain. It has also been shown to help improve immune function, reduce blood pressure and cortisol levels and increase telomerase activity.

What is the underpinning neuroscience?

In general, evidence shows that changes occur in the anterior cingulate cortex, insula, temporo-parietal junction, fronto-limbic network and default mode network structures. Specifically, it appears that mindfulness-based meditation leads to less involvement of the dorsomedial prefrontal cortex (DMPFC) during interoceptive attention (IA). Drawing on some of what we learnt in Chapter 4 about the insula we can now add that mindfulness also leads to altered functional connectivity between the DMPFC and the posterior insula. It leads to increased IA-type activity in anterior dysgranular insula regions, which is consistent with integration of interoceptive sensations with external contexts – or, in other words, appropriate feelings. Overall, we're seeing that this type of interoceptive training modulates task-specific cortical recruitment.

Table 28.1 lays things out clearly. Each of the mechanisms we understand to be key in mindfulness have been seen to correspond with specific areas of the brain.

TABLE 28.1   Key mechanisms in mindfulness as they correspond with specific brain areas

Mechanism	Brain area
Attention regulation	Anterior cingulate cortex
Body awareness	Insula, temporo-parietal junction
Emotional regulation – reappraisal	Dorsal prefrontal cortex
Emotional regulation – exposure, extinction and reconsolidation	Ventro-medial prefrontal cortex, hippocampus, amygdala
Change in perspective on oneself	Medial prefrontal cortex, posterior cingulate cortex, insula, temporo-parietal junction

Long-term meditation has been shown to increase cortical thickness in anterior regions of the brain (including the medial PFC and superior frontal cortex). Conversely, reduced cortical thickness has been found in posterior regions of the brain (including inferior parietal cortex and posterior cingulate cortex). In short, people who have meditated for a long time have structurally different brains to those who haven't.

Being able to focus attention is a very important skill for anyone working in a role where they have to use their brain to think a lot. We know that the anterior cingulate cortex enables this executive attention to take place by acting as a conflict detector. It detects the conflicts between incoming information streams that are being processed and that are incompatible with the task goal. It alerts other parts of the brain to regulate these streams.

Additionally, recent studies have indicated that different types of meditation involve different brain networks or neuroplasticity. For example, the Theravada traditional meditation that focuses on loving kindness has correlated with increased grey matter in the parahippocampal gyri (for five-year-plus meditators). This area of the brain is involved in cognitive empathy, anxiety and mood.

What are any interesting studies in this area?

Many studies have been done over recent years and this has become a very hot topic (See, for example, Lazar *et al*, 2005; Lutz *et al*, 2008). One study looked at whether mindfulness practice had any impact on functional connectivity between default mode network (DMN) regions (the specific set of regions that are engaged when people are left to think to themselves undisturbed, eg remembering the past or envisioning the future). The participants of the study went into an fMRI scanner during a resting state and the data collected showed that experienced meditators, when compared with beginner meditators, had weaker functional connectivity between DMN regions involved in self-referential processing and emotional appraisal. They had increased connectivity between some other regions, notably the dorsomedial prefrontal cortex and right inferior parietal lobule. The conclusions that can be drawn from this is that meditation appears to change functional connectivity in core DMN regions and that this could reflect strengthened present-moment awareness.

A recent study led by Dickenson (2013) looked at the neural mechanisms of focused attention (in this case the practice of focusing on one's breath). The participants had to complete a task while utilizing mindful attention (focused breathing) and control circumstances (unfocused attention) while undergoing a fMRI scan. It was observed that during focused breathing a neural attention network was recruited that included parietal and prefrontal structures. This leads us to believe that attention processes benefit from trait mindfulness. Tang and Posner (2009) showed that five days of integrative body–mind training may lead to greater activation of the rostral ACC during the resting state. Tang (2010) also showed that after 11 hours of this training there was an increase in white matter integrity in the ACC. Other studies have shown that this initial enhanced activation of ACC may later decrease as expertise grows and the ability to focus attention is steady. Electroencephalogram (EEG) data showed increased frontal midline theta rhythms during meditation. This form of brainwave is associated with attention-demanding tasks.

Several studies have documented the stress-reduction effects of mindfulness:

- Brown and Ryan (2003) showed a reduction in mood disturbance and stress following mindfulness interventions.
- Jain et al (2007) highlighted the specific effectiveness of mindfulness versus other relaxation techniques. They suggested that it had a 'unique mechanism by which mindfulness meditation reduces distress'.
- Jha et al (2010) found that meditation training practice could offer protective benefits against high-stress environments.

A specific mindfulness-based stress-reduction programme was created by Jon Kabat-Zinn and the use of this is increasing worldwide.

Studies have also shown that:

- Mindfulness-based interventions (MBI) are effective for reducing depressive symptoms.
- Mindfulness-based stress reduction can reduce emotional reactivity.
- MBI may help people to stop smoking.
- MBI can help executive control, eg emotional acceptance and performance monitoring.

Dispositional mindfulness is an innate characteristic that reflects individuals' naturally occurring ability to inhabit this intentional stance of awareness. Studies have found that people with higher dispositional mindfulness (measured by the Mindful Attention Awareness Scale) predicted increased activation in the ventromedial PFC, medial PFC and ventrolateral PFC. It also predicted reduced amygdala activity and a stronger inhibitory association between the amygdala and PFC.

What can I do with a client this afternoon?

Introducing your clients to mindfulness could benefit them in many ways:

- more success at work – increased ability to focus;
- better health – less stress;
- improved relationships – better emotional control.

We know that people who are mindful enjoy these benefits and we also know that mindfulness appears to be able to be learnt. Referring your clients to a course or book on mindfulness could be something they thank you for. Then for clients who choose to pursue an increase in their meditative capacity, offer support through your coaching programme with them.

Mindfulness may even be something that if you are not yet practising yourself, you choose to look into.

The five facets of mindfulness are:

1 Observing (noticing internal and external stimuli such as sights, sounds, smells, emotions, sensations and thoughts).

2 Describing (giving these stimuli a name – such as 'fear' rather than 'Bob').

3 Acting with awareness (being cognisant of what you are doing rather than allowing your mind to be elsewhere).

4 Non-judgement (not giving a positive or negative value to anything that is happening).

5 Non-reactivity (allowing everything to pass by rather than attention getting stuck in individual thoughts or feelings).

Body scan meditation

This is a simple practice that you could try yourself:

1 Lie down comfortably on your back. Ensure you won't be disturbed and are warm enough. You can either close your eyes or keep them open, or open them during the meditation at any time (especially if you feel yourself falling asleep!).

2 Bring your awareness to your body and the physical sensations you can perceive. Focus on the points where your body is connected with whatever you are lying on. Each time you breath out, feel yourself sinking deeper into that surface.

3 Consider this as an opportunity to 'fall awake', becoming acutely aware of what you are actually experiencing, rather than whatever you think should be. During this time you are not *trying* to feel or be anything, your only task is to increase your awareness of what is. Equally, if you are not feeling anything at any point, physically or otherwise, then that too is absolutely fine and only to be noticed.

4 Take your awareness first to your tummy, noticing the changes as you breath in and out, and any areas that feel different to others.

5 Next gather your attention into a spotlight with two beams and travel it down to your legs, into your feet and out into each toe. Notice what you feel – contact, tingling, warmth, numbness, nothing? Whatever you do or do not feel is whatever it is. There is no need for any judgement to be placed on anything you feel. You can allow any of the sensations to just be whatever they are.

6 As you next breathe in, imagine that inhalation going through your lungs all the way down your body and out into your toes. When breathing out, imagine that breath flowing out of your toes, your feet, your legs, tummy and finally out of your nose. This is a different experience for many people so may take a little playful practice.

7 When you are ready to move your attention (on an out breath) from your toes, bring it to the sensations in the bottom of your feet. Notice each of the different parts of your feet; explore the sensations present there.

8 Allow your awareness to fill the rest of your feet and then, on an out breath, let your feet go and move the attention to your lower legs.

9 As the name implies, you can continue to scan each part of your body one section at a time. Perhaps you might spend 20–30 seconds on each area, maybe a little more or a little less. The method is just to keep your attention there for as long as it feels right and then move on, rather than counting breaths or using a stopwatch!

10 If you find any areas of intense sensation, for example an area of tension or pain, explore breathing into this area. Breathing in, bringing your attention into the sensation and breathing out, noticing if anything changes.

11 Finally, experience the body as a whole, breathing in and out, holding as many of the sensations in your awareness.

Mindfulness exercise

Another straightforward exercise to try is to practise doing any normal activity mindfully. The activity can be anything – cleaning your teeth, doing the washing up, driving a short journey, getting dressed, or anything you do regularly. The aim is not to enjoy the activity more or to do it in a different way. The only aim is to be as *awake* or *alive* to it as possible. Focus your attention on each of the sensations you experience as you do it. Notice any thoughts or feelings that you have as you do it.

Flow

29

What is it?

Being in flow can be described as a specific optimal state. It is a psychological construct and is often called being 'in the zone' or 'on fire'. It can be characterized as engaging in challenges that stretch you, but that you are capable of. Tackling a series of goals, continuously processing feedback about progress and adjusting action based on this feedback are key ways that the state of flow is achieved. Psychologist Mihaly Csikszentmihalyi proposed this state and suggests it is possible to experience it in any area of life: work, relationships, hobbies, sport etc.

Why is it important for coaches?

The state of flow is a highly regarded construct. It is covered here because even if further research uncovers multiple constructs in the future (as apposed to just the state of flow), the 'knowledge to action' gap means that the concept is likely to be around for some years yet. Flow is considered productive, fulfilling, satisfying and positively feeds back into you.

The state of flow has the following characteristics:

- intense, focused concentration on the present task;
- action and awareness are merged;
- low self-consciousness;
- sense of autonomy;
- feeling as if time has passed quickly;
- activity is intrinsically rewarding.

When people are either overstimulated or understimulated their levels of performance are negatively affected. To perform well requires the 'Goldilocks' level of arousal: enough to be energized and focused but not too much so that anxiety and panic freeze your cognitive abilities and ability to act. Today, more so than ever, organizations and cultures have been leaning towards rewarding and validating people who work like crazy. The irony is that it *is* crazy! Pushing and rushing is not conducive to the way the brain works optimally. Too much arousal can lead to high levels of cortisol – and over time this can damage our brains.

What is the underpinning neuroscience?

Currently the neural correlates of flow are not conclusively researched (I'm hopeful that this will change in the near future... certainly researchers realize that there is a gap here). It has been suggested that rapid attentional shifting between salient cognitive precepts correlates to feelings of elation and satisfaction. The neurochemical process that enables this shifting increases cognitive efficiency and creativity. Cognitive shifting has been associated with dopamine release, and subsequently with reinforcement.

As you will remember, dopamine enables us to give our attention to something (it stabilizes active neural representations in the prefrontal cortex); dopamine can give us advanced rewards before a behaviour has even occurred. The way that people describe the state of flow could be linked to the cognitive effects of dopamine. It is conceivable that mesolimbic dopamine facilitates switching between alternative cognitive perspectives and so enhances decision making and creative thinking. Neuroimaging studies have shown an increase in dopamine release during activities that require sustained shifting of a cognitive set, such as playing video games.

There appears to be strong links between the practice of mindfulness and the state of flow; more research is needed, however, to flesh out the details. We do know that neuroplasticity enables us to rewire our brains. It is plausible, although untested, that some brains are currently better wired than others to facilitate the state of flow. For example, some people's brains may be better wired for focusing. When we are focusing well we tend to enjoy what we are doing more and don't experience undesirable ruminating and inefficient distractions. We end up performing better – quicker and with fewer mistakes – when we are focused.

What are any interesting studies in this area?

There is an interesting study by John Pates that looked at the effect of hypnosis on flow states and the performance of basketball players (Pates, Cummings and Maynard, 2002). It was suggested that people enter an altered state of awareness during hypnosis and move from a cognitive and analytical processing state to a more holistic and imagined one. There is some evidence that this is the case and that EEG activity shifts from the left to right hemisphere during hypnosis. Plausible next-stage suggestions are that hypnosis assists in purposeful hemispheric shifts that are conducive to performance, eg control over flow. Pates did a couple of very interesting things that are useful for helping people to access the state of flow. The first is to use trigger-control techniques: anything that can act as a trigger (such as words, sounds, images, components of tasks). The second thing was to encourage, in this case basketball players, to recall and submerge themselves in the memory of a time where they performed optimally. Associating this state with a trigger enabled them to access this optimal performance during a future event. In this study, the players enhanced their three-point shooting performance. It showed that flow can be accessed using hypnotic regression and trigger-control techniques.

Some studies have shown that work environments should have open space, windows and enable people to work without being interrupted. There should be a culture of achievable challenge and a higher purpose in what you are doing. These things increase the likelihood of people getting into the state of flow. For the state of flow to occur, it is important that your skill level matches the challenge. You need to be capable of doing the task.

An experiment from a very different angle involves transcranial direct current stimulation (tDCS). In essence, this is a small electrical current that is directed through specific parts of the brain. Some studies have shown this to produce in individuals a state similar to flow. This electrical current is thought to depolarize the neuronal membranes in the area of the brain that it is being directed to. This makes the cells more excitable and responsive to inputs. It is possible that this process accelerates the formation of new neural pathways when an individual practises a skill. In one experiment using tDCS the subjects were tasked to play a computer game in which they were shooting at targets: the results were that the response speed increased by 2.3 times what it was without the tDCS. How tDCS works is not completely clear yet, but we know some interesting things. For example, tDCS gives people an advantage as soon as the current starts to flow; this leads us to believe that the mechanism is not solely about writing new memories, because the process wouldn't be that quick. A possible explanation links to the theory that activity in the prefrontal cortex is reduced. We know that this area is active during critical thinking, and Csikszentmihalyi has shared that this is reduced when people are in flow. Neuroscientist Roy Hamilton believes that tDCS may have much broader effects than we currently realize. While some neurons are muting signals of other brain cells in their network, it is conceivable that by stimulating one area of the brain, other areas may reduce their activity.

There are several studies that look at flow at work. One study looked at 258 secondary-school teachers. The researchers were exploring the relationship between personal and organizational resources and work-related flow. The hypothesis was that personal resources (for example, self-efficacy beliefs) and organizational resources (for example, social support and understanding clear goals) facilitate work-related flow and also the converse, that work-related flow has a positive influence on personal and organizational resources. Flow at work can be thought of as a short-term peak characterized by absorption, enjoyment and intrinsic motivation. An absorbed employee is in a state of total concentration and experiences time flying while they are immersed in work.

The researchers focused specifically on self-efficacy (our belief in our own ability to achieve something). There is a lot of research into self-efficacy already and it has been shown to be linked to resilience, an increased sense of competence to control and impact one's environment, and work engagement; it is seen to be a buffer to stress and is linked to better health, self-development and social integration. The researchers found that personal resources and organizational resources did indeed facilitate work-related flow.

The experiment made use of the self-assessment WOLF (work-related flow) scale. They also found a causal relationship between work-related flow and a positive effect on personal and organizational resources. Apparently, feeling competent in the present was a good predictor of being in flow in the future.

Another very different study looked not at the individuals and the state of flow but at flow experiences at the group level. It has been previously noted that *collective flow* is more likely to occur when a group is performing at the peak of its abilities. Even more interestingly, a multi-sample study showed that social flow is more enjoyable than the individual flow experience. This fits with other research that we are aware of around the importance of our social nature. According to social cognitive theory (SCT), when people collectively share the belief in their ability to achieve something this may impact how the group perceives challenges and may contribute to the experience of collective flow.

What can I do with a client this afternoon?

- If your clients are not yet familiar with flow you could encourage them to read up on it.
- Ask them to keep a note during a normal day of how many times they think they were in the state of flow, and what specifically they were doing at that time.
- For clients who are not sure when they experience flow, coach them through discovering when they might be in that state (for example during sport or music) or strategize what they could try in order to experience it.
- Consider embarking on a programme addressing any self-efficacy beliefs and helping your client to build and strengthen these.
- Look for ways they can increase opportunities to get into the state of flow every week.

University Centre Library
The Hub at Blackburn College

Customer ID: ******02

Title: Existential perspectives on coaching
ID: BB60026
Due: Thu, 04 May 2017

Title: Business coaching : achieving practical
results through effective engagement
ID: BB65083
Due: Thu, 04 May 2017

Total items: 2
23/03/2017 10:11

Please retain this receipt for your records
Contact Tel. 01254 292165

For clients who are struggling to find this state of flow, a more comprehensive approach may be beneficial. Perhaps start by sharing with them the analogy of thoughts being like trains. The trains may keep coming into the station, but you have the choice of whether or not to get on. Your thoughts may keep coming into your head, but you can decide whether or not to go with the thought or just let it pass out of the station again. Our thoughts can be incredible distracting – and for the state of flow, like with mindfulness, it is important to be present. Strengthening our ability to let our thoughts pass and instead to stay in the moment could make it easier to get into and stay in the state of flow.

Motivation

30

What is it?

Motivation is what drives and sustains a person to act towards a goal. It has movement and direction. This goal may be a conscious and clearly laid out one or it may be more subtle and unconscious. Often the situation is such that motivation is only noticed when there are challenges with getting things done, and it is then that people want to address it. Motivation is a psychological construct, but from a neuroscience perspective it involves the dopamine reward circuitry in the brain. We are driven to undertake activities that are rewarding to us and avoid punishment to us. We form predictions about what will reward us and we use these to form the basis of decisions that guide our behaviour.

Why is it important for coaches?

The ability of a client to achieve a goal is heavily dependent on how they manage their motivation to keep taking the steps required to move towards that goal. If a person is not motivated then they are unlikely to take enough action to make something into a habit or to get the result they want. Motivation is a key concept that coaches need to be aware of. This is true for both people working with individuals on personal aspects of their life and for people working on work-related things.

Imagine a manager who, acting daily in the capacity of a coach, understands the science behind how people are motivated. The power that individual would have to help blend together the elements that would best serve their team would be immense. The knock-on benefits that the team would experience would be great. They are likely to enjoy increased productivity, improved job satisfaction, increased loyalty and a sense of belonging, among others.

Some people ask whether it is possible to motivate another person – and there are challenges inherent within that question. If we think of motivation purely in neurological terms then motivation exists within the individual. However, we are not just our brains – our environment and behaviours can and are shaped by others.

What is the underpinning neuroscience?

There is a lot to motivation, and there is not scope to cover everything here. However, recent reviews of the scientific literature draw distinctions between two types of motivation. One focused on motivation and goal-directed behaviour, the other focused on habitual or stimulus-response learning. Let's start with the habitual or stimulus-response learning. At a simple level, motivation involves the reward circuits in the brain. Dopamine neurons in the mid-brain signal a reward prediction error, which lets us predict – and take actions to increase – the probability of receiving a (neurological) reward in the future. This is the underpinning to lots of habitual learning. For example, if your client does something at work and then their boss says 'well done, that piece of work really helped us secure important new business', then it is likely that your client will be more motivated to repeat the behaviours that led to the praise – and that gave them the dopamine hit.

The reward system in the brain has important links to the area of motivation. Rewards can reinforce behaviour. If we perform a behaviour that is then positively reinforced, the motivation to do it again increases. A reward is considered a reinforcer if it increases the probability of the recurrence of a behaviour. The reward circuits include the dopamine-containing neurons of the ventral tegmental area, the nucleus accumbens and part of the prefrontal cortex.

The reward system uses a major neurochemical pathway, the mesolimbic pathway. This goes from the ventral tegmental area (VTA) via the medial forebrain bundle to the nucleus accumbens (NA) (within the ventral striatum). The star neurotransmitter of the reward system is dopamine and this is released primarily by the VTA. The way dopamine works in this reward system is not yet clear-cut. There are three solid hypotheses:

1 Hedonia – here dopamine is positioned as a 'pleasure neurotransmitter'. When people take certain addictive drugs and experience a 'high', studies have shown dopamine to be involved. However, it has also been found that not all rewards or pleasurable experiences involve dopamine.

2 Learning – in this context learning involves predicting future rewards and forming associations. Studies have shown that rats whose VTA and NA have been destroyed are still able to learn, but are not motivated to work for a reward.

3 Incentive salience – the feeling of wanting something may be a front-runner in the hypotheses. When we want something, dopamine is released in order to help us to work hard to get the thing.

Goal-directed behaviour is different, however, and the underpinnings here are a little less well researched. We do know that the prefrontal cortex is important in goal representation, decision and planning. A lot of the research has been done in rats and there is good evidence that reward learning, grooving in of stimulus-response habits, is under dopaminergic control. The processes going on here, though, seem to be neurally, behaviourally and computationally different from goal-directed actions. It is suggested that there are parallels between goal-directed behaviour and nondeclarative memory (unconscious memories, skills). It is not surprising that decisions that people make involve memories that they have, nor that these cognitive capacities are shared.

While the neuroscience is a little thin on the goal-directed behaviour front, we have a good body of knowledge of habitual actions and the motivation that is involved there. From a coaching perspective we do know that it is important to help clients to develop habits that support them, and memories that help to guide and drive them.

Making decisions

When we are considering our options we need to weigh up which will give us the biggest reward. For example, perhaps for the small-business owner it may be the choice between relaxing on the sofa or working on their new website; for the working parent perhaps it is getting an extra 30 minutes of sleep or getting up and going for a jog. Our brain is responsible for predicting where the highest reward will come from. This process is a complex one and involves balancing exploitation and exploration – exploiting what is already known and coded for in terms of rewards from actions, and exploring new actions to see what the reward is. As individuals our responsibility is to make it easy for our brain to wire in rewarding behaviour repeatedly, so that it makes it easier for us to choose it again in the future. As coaches our responsibility is to support and guide our clients in doing this.

What are any interesting studies in this area?

We'll start with a psychology study led by Sam Glucksberg in 1962. Since the 1940s an experiment called the 'candle problem' had been used, which involves asking a person to fix a lit candle to a wall in a way that the candle wax doesn't drip on to the table. They can use anything in front of them, which includes a book of matches and a box of thumbtacks. To do the task the person has to overcome prior programming telling them that the box is for the thumbtacks. The answer to the problem is to tack the box to the wall and put the candle in the box. What Sam did that was new was to offer participants a financial incentive. One group were told that if they were quick then they would be given some money (enough money to be attractive at that time). What happened to the group who were offered this incentive, though, compared to the group who were not, was that they were *slower*. That's right... the financial incentive made people slower (Glucksberg, 1962). In fact, in such circumstances as this, the more money people are offered, the slower they become!

Another interesting study was done at MIT where they took groups of students and gave them challenges to carry out, including things like solving word puzzles, number puzzles, even physical things such as throwing balls (Pink, 2011). They incentivized the students using a three-tier system. The results were that with mechanical skills the bonuses did increase productivity: the students performed better. However, as soon as even rudimentary cognitive skills were required for tasks, the rewards scuppered people's ability to do them.

Reinforcement learning

One experiment involved a monkey that was irregularly given drops of juice that were signalled by prior visual cues. The researchers found that when the reward (the juice) is unexpected (in this instance if the visual cue is not reinforced or there isn't a cue) then dopamine neurons are excited. On the other hand, when the reward is expected because there has been a reliable cue, then the dopamine neurons do not respond. On a sliding scale, the more expected the reward, the less the neurons respond. When a reward is expected but it fails to materialize then the neurons are briefly inhibited below their baseline-firing rate. So this 'reward prediction error' shows that dopamine neurons fire relative to the baseline to notify the difference between expected and actual reward.

Error signals are useful to us to update our predictions about the future. Evolutionarily it has been advantageous to be efficient. If we can improve our ability to gauge the likelihood of a reward, then we can evaluate how much energy we want to expend on something. If something is not rewarding to us then we are not motivated to do it.

Reward

A study that looked at the responses to reward and punishment in the basal ganglia. Participants in this study were put in fMRI scanners and then had to guess whether the value of a card they were about to be shown was higher or lower than five. They were given a financial reward if they guessed correctly. Different trials gave them reward (+$1), punishment (–50 cents) or neutral feedback. Parts of the dorsal and ventral striatum responded differently to this feedback. After a reward feedback there was sustained activation, whereas it decreased below the baseline after punishment feedback.

A review of findings from neuroimaging studies shows the orbitofrontal cortex to be involved in coding stimulus reward value and working with the amygdala and ventral striatum to represent predicted future reward. In one study, participants were scanned while being presented with predictive cues. These cues were one of two odours. One of the odours was then devalued in the brain of the participant by using selective satiation. This enabled the imagers to see that the brain regions that responded to predictive cues, tracking changes in reward value to the odours, were the orbitofrontal cortex, amygdala and ventral striatum.

Salience – how noticeable or important something is to us

The striatum is implicated in coding for how salient something is, similarly to dopamine. The striatum has been activated in studies where 'non-rewarding' salient events occur. When the involvement of dopamine neurons in rewarding versus salient events is debated then it has been useful to see whether these neurons also respond during salient but punishing events. A similar approach can be taken when looking at the involvement of the striatum and salience versus reward. Interestingly there are studies that show that the striatum *does* respond when an aversive stimuli is present, for example during pain or the anticipation of pain. When someone is expecting a reward, but doesn't get it, parts of the ventral striatum are deactivated. It is important to note that not getting an expected reward is equally if not more salient than getting an unexpected reward.

What can I do with a client this afternoon?

Help your client to identify what already motivates them. This is likely to be causing a release of dopamine. See if you can unpack any obvious links between what is driving them and what is important to them.

People generally respond well to:

- autonomy;
- purpose;
- mastery.

So in your coaching consider focusing in on each of these areas and exploring how your client is getting and working towards each of these things.

As an experiment, listening to your playlist on random or to your favourite radio station may give hits of dopamine as the songs play and surprise you.

For the manager coach

It is really important to pay attention to the effect that expectation has on the brain. We saw from studies that expecting a reward that then does not materialize has an effect. We can liken this to a demotivating effect. On the other hand, an unexpected reward can be likened to a motivating effect (it isn't quite as simple a translation but there is potential for the effect to be as such). Therefore, precautions need to be taken to ensure that people are best set up to take actions that trigger the reward system. This may be as simple as reviewing targets to be achievable or to have positive social interactions within teams. Alternatively, it may be more along the lines of you giving your team unexpected rewards – such as a surprise box of doughnuts, letting them go home early, or words of praise in front of other teams.

Decision making   31

What is it?

Decision making is the process of selecting one choice from multiple options. You can decide to do something, think something or feel something. We make countless decisions every single day and it is only relatively recently that we have been able to look at how this process occurs in the brain.

Why is it important for coaches?

Clients make decisions all the time, mostly unconsciously. Many of their results are the product of the decisions they make. Achieving goals hinges on people being able to make decisions that will lead to the accomplishment of those goals.

Lots of people still believe that decision making is a purely rational process. This gives coaches an opportunity to expand that understanding to take into account the most recent findings in this field.

What is the underpinning neuroscience?

Decision making is a complex cognitive task, which includes many variables. Making a decision can involve accessing memories, a value system and an emotional response. Often we need to accumulate evidence for and against the different choices available to us, evaluate the possible outcomes and risks, and suppress learned responses – and all these processes can be happening unconsciously.

Several parts of the brain have been identified as important to this complex process, including the orbitofrontal cortex (within the prefrontal cortex) and anterior cingulate. The orbitofrontal cortex is responsible for processing, evaluating and filtering information both of a social and emotional nature. The anterior cingulate is responsible for controlling and selecting appropriate behaviour and monitoring errors and incorrect responses. The dorsolateral prefrontal cortex (dlPFC) is involved indirectly in decision making. It is involved in working memory, planning and flexibility.

There are some other interesting things around decision making from a neuroscience perspective. Sometimes we selectively search for evidence of something and disregard evidence for something else (termed confirmation bias in psychology): for example, listing several people who are old and who have always smoked or eaten an unhealthy diet, while ignoring all those who die from diseases related to these activities. People who are very defensive in this kind of decision making have been shown to have increased activity in their left prefrontal cortex.

We can be primed to pay more attention to recent information when compared to things we haven't come across in a while. For example, if the last article you read on exercise endorsed the benefits of interval training then you may do that at the gym tonight rather than your usual distance training.

Only recently has it become possible to do single-neuron recordings that extend beyond sensory and motor coding, beyond what we see or do. It is now possible to study the processes involved in how we value and make choices. It is useful to remember that in the area of decision making our behaviour is very adaptive. Our environment is very important and this is reflected in studies that record the activity of single neurons.

What are any interesting studies in this area?

There are researchers seeking to understand the physical substrates of mental phenomena. Scientists are out there trying to work out how our thoughts are generated by brain processes. When dealing with something as potentially subjective as thoughts, a quantitative approach is valuable. This is something that economic models bring to the field of decision making.

There are a range of types of studies that look into decision making. There are multiple single-neuron recordings that identify which neurons are active during a decision-making task; there are PET scanning studies and functional MRI studies that show increased activity in bigger areas of the brain. The non-invasive scanning experiments are where we focus.

Antonio Damasio has been a pioneer in shaping our understanding of the role of emotions in decision making (Damasio, 1996). Damasio's somatic marker hypothesis suggests that the decision-making process is one that actually depends on emotion. Studies have shown that when the ventromedial prefrontal cortex (vmPFC) is damaged, and the normal emotional signals can't get through, that advantageous decision making is disrupted. Other studies have shown that the amygdala also effects decision making but in a different way to the vmPFC.

Alan Sanfey has enlightened us greatly into social decision-making by drawing on experiments from game theory (Sanfey, 2007), which is a collection of rigorous models that bring to light complex decision-making interactions. He explains that the field of neuroeconomics is integrating cognitive and neural constraints of decision making, bringing together psychology and neuroscience.

Studies using the trust game, prisoner's dilemma and the ultimatum game have all uncovered interesting insights into how our brains work when making decisions. Some of the themes that have come out of this research include social reward, competition, cooperation, coordination and strategic reasoning. It is believed that the mesolimbic dopamine system is involved. Changes in the striatum have been shown to correlate with increasing magnitude of monetary rewards or punishments. It has been suggested that the striatum may register social prediction errors to help guide decisions about reciprocity. One study showed that beliefs about the person they were about to play a game with reduced the amount of 'trial-by-trial' learning required, which showed that the neural basis of social cooperation goes both top-down and bottom-up.

A fMRI study looked at unfair behaviour in the ultimatum game and found that the anterior insula showed greater activation as the unfairness increased. This activation predicted whether or not the player would accept the other player's offer – what decision they would make.

Several studies have highlighted the difference between cultures where individuals tend to be more or less independent. This can have a big impact on decision making. For example, in North America people are more likely to decide to buy sun-tan lotion after watching a promotion-based advert, while on the other hand, East Asians will be more likely to buy sun-tan lotion after watching a prevention-based advert. This only holds true when the people are time-pressured, though: this means that if you are working with someone who has to make quick decisions then looking at what their biases are, and how they are predisposed to make decisions, could be useful to them.

Another interesting component to decision making is the phenomenon of *Schadenfreude*. This is where people 'delight in others' misfortune'. From a social comparison theory perspective, when people around us are in pain in some way (socially or otherwise) we look better to ourselves. In one MRI experiment that was designed to measure empathy they found that men (but not women) enjoyed seeing bad people suffer. The experiment involved participants watching 'bad' or 'good' people being given electric shocks. The researchers expected to see the empathy centre of the brain light up, which it did, but in men so did the pleasure centres when they thought that the subject was deserving of punishment.

There are links between *Schadenfreude* and envy: one study could predict the brain's *Schadenfreude* response based on the previous envy response; a person's ventral striatum (reward centre) was activated when they were told that people they envied had suffered misfortune.

What can I do with a client this afternoon?

Task clients with reflecting on some of their best decisions over the last year. Which ones do they feel really good about? Reverse engineer how they believe they made these decisions and look for any patterns. Are they aware of any intuitive decision-making component? Do they tend to need the logical arguments to stack up before they can commit to something? Have they ever had the experience of making a decision and then feeling like it was a good one once it was made?

Explore with a client the value of emotions in the decision-making process.

Expectations

What are they?

An expectation is what we think or hope is most likely to happen. We have expectations all the time; it helps us manage our world. When an expectation is unrealistic, though, there is a high risk that it will not be met.

Why are they important for coaches?

Everyone has expectations that they may or may not be consciously aware of. When expectations are not met then it can have an adverse effect on a person. When they are met or exceeded it can have a very positive effect. Understanding these and how to use them to your advantage can be useful. In the workplace, often the expectations of others also need to be managed for a smooth relationship.

What is the underpinning neuroscience?

Expecting something positive to happen can have the effect of reducing anxiety. This can be useful if, for example, a client has a meeting coming up and they want to feel calm about it. They may want to imagine it going well and them conveying their points effectively during this meeting. It can also activate the brain circuits that are involved in reward mechanisms. These reward circuits link together both cognitive and emotional responses, and even motor responses. This may mean that someone enjoys the benefits of a reward circuit activation by just thinking about a positive expectation they have.

On the contrary, having a negative expectation can increase anxiety and activate the areas of the brain involved with negative experiences. This obviously affects our state and subsequently has the potential to affect the end result.

So, when we expect a primary reward, dopamine cells seated in the nucleus accumbens fire off in anticipation. When we are not expecting a reward but receive one (such as a surprise compliment, a hug, finding a £5 note on the floor, etc) we get an even bigger dopamine hit. When our expectations are unmet – like when we don't meet a deadline, gain weight we'd expected to lose, don't get that promotion we'd been promised – the fall in dopamine levels is painful. These and a huge number of other unmet expectations can also cause a mild threat response.

As you'll remember, dopamine is really important for us in order to be open, curious and interested. How fast we can think is increased with more dopamine – too much, though, and our minds become chaotic. Being able to hold an idea and focus in on something is dependent on the right levels of dopamine.

What are any interesting studies in this area?

Scott *et al* (2007) did an imaging study that looked at the role of the nucleus accumbens and expectations. They took subjects and used functional molecular imaging to observe the dopamine release when people were being given a placebo. They found that the activation was related to the anticipated effects of the placebo! They then did some functional MRI studies, which looked at the expectation of monetary gain. Here they found that this expectation increased the nucleus accumbens synaptic activity proportional to the dopamine release when people were given placebos.

Another study, this time by Volkow *et al* (2004), looked at the effect of expectations and the reinforcing effects they could have. In this study, cocaine users were split into four groups:

1 People expecting a placebo and given a placebo.
2 People expecting a placebo and receiving the drug methylphenidate (some similarities to cocaine).
3 People expecting the drug and receiving the drug.
4 People expecting the drug and receiving the placebo.

The results were significant. When people expected the drug their brain glucose metabolism was 50 per cent higher than when they didn't expect the drug. People also self-reported feeling high by 50 per cent more.

When people are given two identical bottles of wine, with wildly different price tags, they tend to say that the more expensive one tastes better! (This gives us good reason to leave the price tag on a good bottle!)

Expectations have also been shown to have a powerful relationship with pain and emotion. We have seen expectancies affect the processing of pain in the insula, cingulate cortex and thalamus. In placebo studies, if we expect something that we are swallowing to reduce our pain, it often does. In another study (for which the details can be confusing, so here we will focus on the results) researchers discovered that our powerful cognitive bias reflects a general property of sensory information processing. In the experiment they test whether the contents of visual awareness could be altered by manipulating expectations. The effects were dramatic and showed that expectations have a strong influence on our experience of sensory input.

What can I do with a client this afternoon?

- Work with your clients to evaluate their expectations and whether they are serving them or hindering them. Mindfulness practices may help some clients to become more aware of their expectations.
- If you coach leaders then help them to examine what expectations their teams have and what affect that may be having on them.
- A classic situation in many business environments is having unrealistic targets that switch people off before they even begin – help clients to address this if it is an issue for them.
- Ensure that clients have planned things to look forward to.

Beliefs

What are they?

Beliefs are tricky from a neuroscience perspective, and in another 10 years we are likely to be in a much clearer place. Neuroscience will have advanced even further and the construct of beliefs will not need to be relied upon so heavily. When we look at commonly accepted definitions we see things like: 'A belief is an acceptance that something exists or is true, especially one without proof.' Or 'trust, faith or confidence in something'. We have a gut feel for what our beliefs are and what we would say are facts that we hold to be true. We can recognize that different people hold different beliefs. In coaching we often say that beliefs can be powerful and have significant effects on people. The concept of beliefs normally can be considered a psychological one.

Eliminative materialism, supported by neurophilosophers Paul and Patricia Churchland, suggests that our common-sense understanding of things, such as beliefs, will one day be proved false. In simple terms, they do not believe that beliefs exist! Instead they propose that neuroscience will uncover a deeper level of understanding that renders the concept of beliefs no longer useful. B F Skinner, a radical behaviourist, was also a supporter of this way of thinking and likened pseudoscientific theories (eg the four humours and phlogiston theory of combustion) to mental states like beliefs.

The Churchlands argue that folk psychology should be evaluated based on its ability to predict and explain things in researching the mind and brain. Neuroscience is able to do this and can explain many cognitive processes that folk psychology is still unable to. Daniel Dennett and Lynne Rudder Baker, who suggest that while beliefs are not a scientifically valid concept they can be useful as a predictive device, propose a more middle ground. If we play chess with a computer, most of us don't think that the computer *believes* that by taking our queen it will have a good advantage. However, we would be sensible to act as if that were the case because it will likely inform the moves that the computer makes. Dennett suggests that beliefs occupy a different level of explanation to those of fundamental neuroscience. Perhaps this is a little like trying to explain a number using colour.

Why are they important for coaches?

Many coaches work with their client's beliefs – and so understanding what they really are, how they are formed, how new ones can be formed and being open to the future offerings of other constructs may be useful to you.

What is the underpinning neuroscience?

Beliefs are relatively poorly defined from a scientific perspective. This causes some problems when trying to identify the underpinning neuroscience. What follows is a range of different approaches that have been explored. Damasio gives us some insight into the complexity of the task when he states: 'it is likely that the mechanisms which allow us to develop the basis of beliefs, as well as the mechanisms by which we retrieve and express them, are operated in a largely covert manner' (Damasio, 2000).

Often in science we learn more about what is normal by examining the abnormal. In this case many people have explored delusions, an impaired perception or belief, in order to better understand what constitutes a normal one and how it may form. However, delusions themselves are also poorly defined. The American Psychiatric Association's handbook (1994) states that a delusion is:

> A false belief based on incorrect inference about external reality that is firmly sustained despite what almost everybody else believes and despite what constitutes incontrovertible and obvious proof or evidence to the contrary. The belief is not one ordinarily accepted by other members of the person's culture or subculture (eg, it is not an article of religious faith).

There have been some correlations between specific delusions and anatomical findings, for example in sufferers of Capgras delusion in which people believe that someone close to them has been replaced by an imposter. A significant proportion of Capgras delusion sufferers have been found to have lesions in the right hemispheres of their brains. Unfortunately, research in this area is not yet producing findings clear enough to be of real use to coaches.

Kathleen Taylor from Oxford University suggests that beliefs are like memories from a neural standpoint. She has written a book that looks at the extreme end of beliefs: brainwashing (Taylor, 2006). Taylor looks first at the basics of neurological pathways, including the communication of neurons and how pathways get grooved in, realizing that the more something is repeated the stronger the neural pathway becomes, which we know is summarized in Hebb's law. When people have strong beliefs it is likely that the neural pathways are strong, perhaps from repetition or intensity. From looking at the techniques employed by cults Taylor shares insights into the importance of repetition and the emotional pull of ideas and new beliefs.

The concept of beliefs is alien to children until they are around four or five years old. If you tell a three-year-old child a story about Jack and his magic bean that he put on the kitchen table, but his mother then moved to the porch, and then ask the child where Jack *believes* the bean to be... most will reply on the porch.

A study that looked at the brains of committed Christians and non-believers gives us some insight into beliefs. The individuals were positioned in functional magnetic resonance imaging (fMRI) scanners and then asked to evaluate the truth and falsity of religious and non-religious propositions. What they found was that when people believed something there was activation in the ventromedial prefrontal cortex. We know this area is linked to self-representation, emotion, reward and goal-driven behaviour. This activation was not dependent on whether the thing that was being believed was a 'Christian' fact (such as the Virgin Birth, for example) or an 'ordinary' fact. When there was a comparison made, they observed that the 'ordinary' facts were more reliant upon memory retrieval networks than the religious beliefs.

What are any interesting studies in this area?

One interesting study that has been linked to beliefs was carried out by Matthew Lieberman *et al* (2005). The study involved both African-American and Caucasian-American individuals who were asked to lie inside fMRI scanners; they were then shown three images of African-American and Caucasian-American people on a screen (one at the top and two below). They had to match the race of the person at the top to one of the two at the bottom. They also had to complete an experiment where they were given just one face and two written words, one of which described the race of the person in the picture (either Caucasian or African-American).

The study was designed to look at any differences between amygdala responses in African-American and Caucasian-American individuals to African-American and Caucasian-American faces. The amygdala is known for its activation to threatening, novel or highly arousing stimulus. It has been implicated in race-related processes too. The results of the experiment showed that both African-Americans and Caucasian-Americans produced greater right amygdala activation to the African-American target than to the Caucasian-American. Previous studies only tested Caucasian-Americans, which meant that the reason for any amygdala activation could not be attributed to a potential threat or the novelty of the African-American face. By this study including African-Americans as test subjects their brain responses can be compared. The authors of the study note that this is not conclusive, although suggest that the observed amygdala activity may be a 'reflection of culturally learned negative associations regarding African-American individuals'.

The other interesting finding supported other studies in the difference between perceptual encoding and verbal labelling effects on the brain. Here, when people saw the words rather than the pictures their right ventro lateral prefrontal cortex was activated. The theory is that this inhibits the amygdala. Some beliefs will light up certain areas of the brain. Many of our beliefs (if we use this term) are picked up unconsciously and we may not even be aware of some that are having an effect on us.

Joel Winston *et al* (2002) carried out another study that asked people to rate the trustworthiness of each of the faces they were presented with. We can think of this as: 'How trustworthy do you believe this person is?' Winston found that the amygdalas in the subjects responded most to the pictures of the faces that had been chosen to represent untrustworthiness. People's ratings didn't always correspond with their brain response. Does this mean that we should trust their brain or their verbal account? What implications does this have for how we work with clients?

What can I do with a client this afternoon?

The suggestions here are mostly familiar to coaches already. Unfortunately at this point we are waiting for science to move on a little more in order to see whether beliefs as a construct will survive another 10 years or whether our understanding will develop. It might be that as a coach you choose to look at and work with the neuroscience in the meantime. Here are some considerations:

- Acting as if beliefs are real:
  - Explain at least the neural mechanism of how thoughts, ideas and possibly beliefs get strengthened (through repetition and emotional intensity and the relevance of Hebb's law).
  - False beliefs are possible in individuals. Therefore, a worthwhile exercise could be to challenge and explore any potential false beliefs.
- The alternative, bold opportunity is to explore how things would be if beliefs did not exist:
  - Explore with your client what they know to be true and how they know it to be true.
  - Explore what your client may now be free to do or be.

PART SEVEN
Neuroscience of not-so-classic coaching areas

Part Seven looks at some tools and concepts that are not traditionally used by coaches or necessarily consciously addressed by coaches. Something like *trust* is, of course, very important in every coaching relationship but is perhaps not paid attention to in other relationships unless there is a problem. Here we get to see some of the far-reaching effects that trusting and being trusted can have on people. The opportunity then becomes one of examining whether actions to increase trust or trustworthiness would benefit the individual.

In this part of the book it is even more important to remember that we are focusing on the definition of coaching as facilitating 'self-directed neuroplasticity'. Like *trust*, things such as *fairness* or *loneliness* do not tend to be focused on unless there is a problem. Coaching is often employed to improve things rather than just as a remedial solution, therefore there is a clear chance here to really add to a coach's toolbox.

The areas of nudging and priming have their roots in psychology. They are very useful for coaches to be aware of to help effect behaviour change without needing to rely on willpower. Leaders and managers who are aware of these techniques have a real advantage.

Choice architecture

34

What is it?

Choice architecture is the process of affecting outcomes by influencing decisions. Typically this is done by presenting choices in a specific way or by making changes to the environment within which someone exists. There are many tools in a choice architect belt, we will look at just two here. We run Behaviour Change training for organizations with our colleagues at the Wales Centre for Behaviour Change. If this piques your interest, you'll find a white paper on this topic at **www.synapticpotential.com**.

Nudging has been described as getting people to make decisions based on how their options are presented. According to the iNudgeyou network:

> The following conditions are necessary (and perhaps sufficient?) of X to be a nudge within a certain choice architecture:
>
> 1 X alters the decision maker's behaviour in a predictable way
>
> 2 X does not limit the choice set, ie the available options to the decision maker
>
> 3 X does not significantly alter the decision maker's incentives, in terms of payoffs and utility increase from the perspective of a rational agent.

X can be a huge range of things. Singapore is a country where nudges are all around you. A beautiful example is where stairs in the Bugis MRT station are brightly coloured. They also say various things like 'I want to climb the stairs to fitness.' There are escalators right next to the stairs, but when they are not being used they are still and quiet. The result is that many people choose to take the stairs.

Nudging finds its home within the field of behavioural economics. This is where the fascinating realization that human decision making is not a purely rational process is explored:

> Nudges are ways of influencing choice without limiting the choice set or making alternatives appreciably more costly in terms of time, trouble, social sanctions, and so forth. They are called for because of flaws in individual decision-making, and they work by making use of those flaws.
>
> (Hausman and Welch, 2009: 126)

Nudging is a big and exciting field. There are specialists experienced in designing nudges, gathering and evaluating data and redesigning nudges. For leaders and managers of organizations who are looking to implement wide-scale nudges that have a lot riding on the results it would be wise to consult the experts. For smaller-scale interventions as a coach you may choose to experiment.

Priming is the second tool we will look at. We can think of the process as activating certain neural circuits and setting us up to respond in a certain way. Technically it is an implicit memory effect where a stimulus influences you.

Why is it important for coaches?

Nudging is important in several ways to your clients. First, as a coach you can suggest things to them, or ask them questions to direct their attention to doing something that will serve as a nudge to them. For example, if you were talking about going to the gym after work, a simple nudge would be to change into exercise clothes at work before getting into the car. This would not limit the individual's choices, they could still drive home if they wanted, but it is likely to result in more trips to the gym. You wouldn't need to explain all the background to a client every time you suggest a nudge. You could even ask questions to help them come up with a nudge, depending on your chosen coaching style.

Second, you could filter for ways that those you work with have unhelpful nudges set up. A leader or manager in a coaching role may see or hear things that are not creating the best environment for individuals. Perhaps people want to be fit and healthy, which is obviously great for productivity, low sick rates, etc. Maybe the layout of the building is such that most people have to walk a long way to take the stairs between floors whereas lifts are right there and perceived as quick. Remodelling the office may not be a viable option, but putting up a sign that says something like 'Treat yourself to thinking time, take the stairs' might result in more people doing so. The exciting thing with nudges is that you can trial them and tweak them.

Alternative ways that you can use nudging is to support those you work with in achieving their goals. So would a motivational e-mail once a week or sporadically help your client? Would a reminder – one that they can print out and put in their work space – help them?

Priming is occurring unconsciously many times during each of your client's days (and yours!). Becoming aware of it and both cushioning against unhelpful primes and utilizing potentially helpful ones is sensible. The power and effectiveness of priming could do with more research, so this is not yet set in stone from a scientific perspective.

What is the underpinning neuroscience?

Nudging involves bypassing the cognitive system; it sidesteps any issues with reactance and attitude. You don't have to persuade anyone to do anything, simply communicate directly with the systems that control their behaviour. As coaches we are extremely familiar with clients' difficulties in changing their habits. The desire to change a behaviour is often met with challenge when the old neural systems that control the habitual behaviour continue to be 'switched on'.

In essence we can think of there being two systems that process different types of information for different aspects of behavioural control. One is the effortful and intentional, while the other is implicit and routine. Evolutionarily this enables us to perform routine behaviours unconsciously, thus freeing up our cognitive system to process other things.

In recent times there have been a lot of questions around priming. Several studies that were done have not been replicated, which casts doubt upon their credibility. The critics have suggested that there has been publication bias, the experimenter effect and that criticisms have not been dealt with well.

There is a little neuroscience that can help us out, although not much yet. In general it makes sense that activating a neural circuit in some way using a prime could make it easier for a linked circuit to be activated shortly afterwards. A study showed that conceptual priming is linked to reduced blood flow in the left prefrontal cortex, which is thought to be involved in the semantic processing of words. Another study suggested that perceptual priming is controlled by the cerebral cortex (specifically the extrastriate cortex). This study showed that perception is improved by prior experience of a stimulus. There was a reduction of gamma frequency oscillations... the assumption is that there are two stages to a prime's effectiveness in the visual system: 1) representation sharpening in the early visual areas; 2) competitive interaction between representations in the higher visual areas and the prefrontal cortex.

What are any interesting studies in this area?

Many of the nudging studies do not have obvious applications to coaching, rather it is the process of nudging itself that you can then teach or apply imaginatively with clients.

A classic experiment involved placing a sticker of a fly in urinals in Schiphol Airport. This Amsterdam airport managed to reduce the spillage, due to bad aiming, by 80 per cent. The UK government had taken notice of this field and had formed a 'Behavioural Insights Team' dedicated to implementing ways to nudge people (that has now become a private entity). Their work resulted in changes to forms and processes in the area of tax collection. This resulted in a 15 per cent boost to tax repayment rates. They also changed the UK driving licence forms to make it 'required-choice' for people to take a stance on organ donation (opting either in or out). This doubled the number of people who are joining the register.

A great series of studies in 2005 by Randy Garner looked at how to increase response rates of correspondence. One experiment looked at what effect it would have if a Post-it note was attached to a survey that was sent to faculty staff at a university. Their aim was to get more staff to fill in and return the survey. They did indeed find that by putting a hand-written request on a Post-it note for individuals to complete the survey on, the response rate was significantly higher than when compared with either handwriting a request directly on to the front of the survey, or attaching a blank Post-it note (Garner, 2005). So the combination of Post-it note *and* hand-written request was the key here.

A favourite study is one that helps people to be more honest. Lisa Shu led a team that looked at what effect was had by getting people to sign at the top of a form declaring their honesty. Participants were told to report on their earnings from a problem-solving task and were also told that travel expenses would be reimbursed and they would be compensated for travel time. The results were stark. Of those who did not sign any declaration of honesty 79 per cent cheated. Of those who signed at the end of the form 63 per cent cheated. From the group of people who signed at the top of the form only 37 per cent made false claims. The reasoning behind this could be as simple as bringing ethics and one's own identity to people's attention before taking some form of action.

One final type of study shows the importance of placement of food. Several variations of this study exist now, and many more could be on their way soon. The study involved rearranging buffet tables to nudge people into eating more apples and fewer brownies. Figures in the range of 80 per cent increase in apples and 30 per cent decrease in brownies are not uncommon in this type of experiment. The subtle changes in our environment really do lead to different behaviours.

In a classic priming experiment John Bargh and his team asked people to assemble a four-word sentence from a selection of five words (Bargh *et al*, 1996). For one of the groups the words included ones associated with the elderly: for example, Florida, forgetful, bald, wrinkle or grey. The other group's word selection didn't contain these words. After each individual had finished the task they had to walk down the corridor to another room to do another experiment. The actual experiment was them walking down the corridor: the time it took each of them to do the walk was measured (covertly) and what they found was that those primed with words associated with the elderly walked slower! This has become known as the 'Florida effect'. When the students were later questioned about their experience of the first task, none of them reported noticing a common theme among the words and all insisted that they couldn't have been influenced by the words. They weren't consciously aware of acting more like an old person!

This ideomotor affect works both ways: so walking slowly makes you more likely to recognize words linked to old age. A common task to lift your spirits is to hold a pencil between your teeth, parallel to your face, forcing you to smile. When students who were doing this were asked to rate cartoons, those who had the pencil this way round rated them as funnier. Other students, who had the pencil held at a 90-degree angle, forcing them to frown, found them less funny.

One of my favourite priming experiments that I cover in *Make Your Brain Work* is what I call 'How to win a game of trivial pursuits.' Two scientists did an experiment to investigate a link between stereotypes priming on overt behaviour. This involved priming people with the stereotype of a professor, a secretary and no priming at all. In this case the priming was done by getting people to read something and there were either strong mentions of a professor or a secretary or neither. The idea was that, stereotypically, professors are thought of as knowledgeable and intelligent whereas secretaries are not. The people were then asked to answer multiple-choice questions – from the game Trivial Pursuit.

The professor-primed people answered more questions correctly than the other two groups, and the secretary-primed people answered the questions most quickly.

What can I do with a client this afternoon?

A great process is to sit down with a client and focus first on helping them identify the top three behaviours they would like to change. Look at what they are doing, when, why, how and make sure you understand what the triggers are to that happening. Elicit what the desired behaviour is instead. Explore what potential nudges could be trialled to unconsciously make the desired option more favourable. Either do this by yourself and make suggestions on what could be trialled for two weeks before evaluating the results, or brainstorm with your client openly. There is less research to show how effective nudges are when people are aware they are being nudged. This is worth bearing in mind but, as with everything here, experimentation is your friend!

Sometimes brainstorming with your client may result in a proposed nudge to help change other people's behaviour, which will then help your client achieve their goals. A classic example is: if a client is finding they are achieving less than they want to in a given day and can attribute this to frequent interruptions then they might choose to close the door to their office at certain times during the day. Very simple, yet often very effective.

On the priming front, challenge your client to open up their awareness for two hours a day for three days, and look at what comes into their consciousness. They won't notice everything (I'm not even sure that is possible!) but they may observe things like:

- When I watch the news, my mood seems lower afterwards.
- When I listen to upbeat music on my way home I work out harder on the treadmill.
- When I smile lots and talk calmly to my children they seem to be easier to look after.

Then ask your client to suggest anything they could put in place that may help them to be in certain states or act a certain way. For example:

- motivational words around their desk;
- walking briskly before a meeting;
- adding some specific pictures around their home and work areas.

There currently is not any direct evidence that this will work... but logically it shouldn't do any harm.

False memory

What is it?

We tend to think of memory as involving the encoding, storing and retrieval of information. There are several types of memory: sensory, short, long, declarative and procedural. While some of our experience of memory is conscious, most of what our memory is involved in is in the unconscious.

A false memory is a psychological term for a memory that might be a fabrication or distortion of an event – or a detail within an event – that didn't happen. It is also conceivable that emotions or feelings may be falsely recollected. Some factors that are known to influence false memory include giving misinformation, existing knowledge and other memories. It is possible to change existing memories to incorporate new information.

Why is it important for coaches?

People may strongly believe that if they remember something this means that it happened. Often problems arise when people act as if their memory is infallible. In a national survey in the United States people were asked if they agreed or disagreed with the following statement: 'Once you have experienced an event and formed a memory of it, that memory does not change.' Nearly 48 per cent of people agreed. It is quite widely believed that stressful events increase our ability to remember something. For example, someone tells their spouse that they want a divorce and a heated interaction ensues. Or at work someone gets a particularly bad performance review, which upsets them. It is not uncommon for people in high-stress situations to really believe that something happened when in fact it may not have done. Based on a lot of research we know that this may not be the case. We know that:

- Memories can be implanted.
- Powerful emotions can reinforce or weaken real memories.
- We can reduce the power of painful memories.
- False memories can elicit the same strong emotions as real memories.

The clients you work with do not exist in isolation. Colleagues, friends, family members all interact with our clients and have the potential to influence their memories. The questions they ask, the summaries they give, the flippant comments they make can all shape what is remembered. You as a coach are also hugely powerful in your potential to affect how your clients remember things.

In a corporate environment, as in life, the stakes can be high. The research that tells us that the presentation of false information can influence what people recall in the future is very important. In a meeting, imagine that someone reports on the last month's events in his team. If he speaks of a conversation the group had last month and says he had mentioned some important data, perhaps believing he had shared this, it has an effect on the group. You are likely to then get division within the group. Maybe 40 per cent of the group will 'remember' him sharing this information, while the other 60 per cent will not.

Acting as if memories are 100 per cent reliable can lead to miscommunication and breakdowns in relationships. The consequences of this can be serious.

In an experiment done with children (whose parents had consented to the study) a man went to a playground and asked individual children to help him look for his lost dog (*Daybreak's Stranger Danger*, 2013). The children were shown a picture of a dog and a lead. Many of them walked out of the playground with this man. Later, some of the children when asked why they did it replied that the man had lost a dog. When pressed, several children said they had actually seen the dog. Interestingly with children, some people's first thought is that they are making up stories rather than they are experiencing a false memory.

What is the underpinning neuroscience?

The neuroscience of memories is quite complex. What follows are some of the highlights of what we think is happening. When you retrieve memory, information is being processed from two different regions of the brain: the medial temporal lobe (MTL) and the frontal parietal network (FPN). The MTL focuses on the specificities of the event while the FPN processes the overall gist of the event. A study using fMRI by Roberto Cabeza showed that when people recalled a true memory they had increased activity in their MTL, whereas when the memory wasn't true (but they still believed it to be) their FPN experienced increased activity.

What are any interesting studies in this area?

Elizabeth Loftus is the woman to watch in this field. She has carried out lots of studies into memory and is known as an expert in the field (see, for example, Loftus *et al*, 2002, 2013). One of her famous experiments is about implanting the memory of being lost in a shopping mall around the age of five years old. This was important because it was the implantation, the creation, of an entirely false memory rather than just a detail within a memory. The way they did it was to initially ask the study participants to try to remember childhood events. They created individual booklets for them with three one-paragraph stories about events that had happened to them and one story that hadn't. The false story was adapted or based on a real shopping trip that the individual had been on. Once the participant had read each story they were to write down what they could remember about the event, or write 'I do not remember this' if that was the case. Two follow-up interviews occurred next, where they were told that the level of detail they could remember, in comparison to a relative, was being examined: 68 per cent of the real events were remembered and 29 per cent of the false events were remembered. This study showed that it was possible, and relatively easy, to implant a false memory.

Another, almost comical, experiment was created in part to prove to critics that the events being falsely remembered hadn't actually happened (making it a retrieved memory instead). For this to work the thing that people were falsely remembering would have to be impossible. The researchers showed people an advertisement that described a visit to Disneyland and meeting and shaking hands with Bugs Bunny: 25–35 per cent of people claimed they had indeed met Bugs Bunny at Disneyland. Of these people 62 per cent remembered shaking his hand and 46 per cent remembered hugging him! Of course this is not possible because Bugs Bunny is a Warner Brothers character and not a Disney one (Loftus *et al*, 2002).

In a very different set of research, subjects were recruited from the US military who were enrolled in Survival School training. This is a programme that is designed to prepare individuals for the possibility of being captured as a prisoner of war. The training is stressful, including being hooded, spending time in a very uncomfortable cell and interrogation that is physically abusive. When they are rescued the soldiers are then questioned about who interrogated them and their experience, as would be similar to what would happen in real life.

During this interview process the soldiers were shown a photo of a person identified to them as the person conducting their interrogation. They were asked questions such as: 'Did your interrogator give you anything to eat? Did he give you a blanket?' In fact, the photo was not of the interrogator at all. Later, though, when asked to identify the interrogator 84 per cent of the soldiers identified the person whose photo they were shown. Similarly other information could be implanted into their memory: 27 per cent of them claimed to have seen a weapon that had not been there (after being fed misinformation); 90 per cent recalled a non-existent telephone. This phenomenon occurred in people who are highly trained to deal with stressful conditions.

By this point you may be wondering if anyone is immune from experiencing false memories. To date, researchers have not found anyone. A study at the University of California conducted an experiment with people who have a highly superior autobiographical memory (HSAM) (Patihis *et al*, 2013). This means that when quizzed these people could remember things like 'On what date did an Iraqi journalist hurl two shoes at President Bush?' or 'What public event occurred on 11 October 2002?' They were also able, when given a random computer-generated date, to say which day of the week this was and to recall a personal experience and a public event that occurred then. The public event was verified using a search engine.

It is amazing to think that someone who can remember that on 19 October 1987 there was a big stock market crash and the cellist Jacqueline du Pré died could have flaws in their memory. But even these HSAM people were susceptible to false memory implants. In one experiment the terrorist attacks of 11 September were talked about and footage of the United Flight 93 crashing in Pennsylvania was mentioned. Later, around 20 per cent of people recalled seeing this footage (both normal and HSAM people recalled this). One HSAM subject recalled: 'It just seemed like something was falling out of the sky.' In various other tests the HSAM and normal memory subjects experienced a similar level of memory distortions.

Bryan Strange from UCL demonstrated the power of emotion in relation to memory in a very simple experiment. His team showed that people were more likely to remember emotionally powerful words such as 'murder' or 'scream' than neutral words. The words that were forgotten most were those shown just before the emotionally charged ones. In this particular experiment the effect was greater in women.

What can I do with a client this afternoon?

- Teach your clients about false memories – empower them to know that what they remember isn't always what occurred, so others may have different recollections – and these also may be distorted!

- Coach them through how they could explain this to others so they too may benefit from understanding how memories actually work.

The real value from understanding false memories is not likely to occur for you this very afternoon. The opportunities to be aware of this could spring up at any time with people you work with. Being able to offer a reason why people remember different things or what may underpin disputes in relationships is when this knowledge could help you to be even more effective.

Trust

What is it?

Trust can be defined as a strong belief in the reliability of someone or something. It can have both emotional and logical components to it. It is possible to get a gut feel that you can trust someone, what they say and will do. It is also possible to rationalize why you should or should not trust someone.

Why is it important for coaches?

Stephen Covey makes some great points about trust and these are important for both personal and business coaches:

> When trust is low, in a company or in a relationship, it places a hidden 'tax' on every transaction: every communication, every interaction, every strategy, every decision is taxed, bringing speed down and sending costs up. My experience is that significant distrust doubles the cost of doing business and triples the time it takes to get things done.
>
> (Covey, 2014)

Most of us experientially know that trust is important. What neuroscience has done, again, is help us to understand how it is important from a very basic level. It adds another piece of the puzzle and enables us to organize ourselves and our workplaces with confidence.

In an economy and world where people are often encouraged to be independent and succeed, even to the detriment of others, this research and understanding is important. As coaches we need to be equipped with as much quality information about how we work as possible so that we can operate beyond cultural changes, fads and ungrounded ideas. Incidentally, Paul Zak (2012) found that the countries with the highest trust levels are also the most economically successful.

Of course the other vital reason for coaches to be aware of the importance and effects of trust is so they can choose to invest appropriately in being trustworthy and building trust with those they work with. In corporate environments where external coaches may have people other than their client to report to this obviously needs to be handled thoughtfully. Most people who start to work with someone in a coaching role – be it a life coach or business-coach-type relationship, or more of a manager as coach scenarios are opening up – could experience vulnerability. The level of trust you can build from the outset can serve you well as the relationship progresses.

What is the underpinning neuroscience?

When we trust someone there is a knock-on effect on specific areas of the brain. The amygdala's activity levels are decreased. Threat and fear responses are low. On the other hand, if trust is breached then the anterior cingulate cortex detects conflict and activates the amygdala. We know that when there is threat and fear this obscures the brain's ability to do other things. Our ability to plan, to make quality decisions and to be creative all take a nosedive. Conversely, when trust is present our brain is freed up to engage in these things competently again.

Paul Zak has carried out a lot of research into oxytocin, which is known to be important in trust. His research, along with that of other great neuroscientists, helps us to understand how our brains process whether we trust or do not trust someone. Oxytocin is known to be important in being trustworthy, in choosing to trust others and feeling trusted. When we feel trusted we can actually become more trustworthy as a result of increased oxytocin levels.

It is worth noting that high levels of stress blocks oxytocin release. Evolutionarily this makes sense. If you are in a stressful situation your attention and focus needs to be on surviving. Oxytocin is important in being empathetic, relating to others needs, and also has an effect on the area of the brain called the amygdala, which would diminish our ability to register anxiety. On a practical level this may mean waiting until someone calms down a little bit before offering them that hug. It also could be the reason that people can feel disconnected during times of stress, and why we may need to work extra hard to connect with them.

Another thing to be aware of when we're working to increase trust by increasing levels of oxytocin is that it can increase emotional pain. Oxytocin can strengthen social memories, meaning that a stressful situation can have a longer-term impact, triggering fear and anxiety in the future. This doesn't mean that building trustworthy relationships isn't a good idea (far from it). It is another component to be aware of, however. The mode of action here, the way that oxytocin intensifies stressful memories (the activation of the lateral septum), leads us to suspect that the same could be the case for strongly positive situations.

If we trust people unconditionally then the area of the brain called the septal area appears to be involved. However, if we trust someone based on certain conditions then the ventral tegmental area (involved in reward circuitry) is activated. This would be a scenario like someone arriving on time for an appointment, sending you a report you've asked for, or remembering to pick up dinner.

What are any interesting studies in this area?

The trust game is a classic research tool that has been played many times all over the world and studied by many different disciplines. Many studies involve the use of it, so an appreciation of the core game is useful.

The way the game works is that a subject sits at a computer screen and it is confirmed to them that they have received a sum of money just for showing up at the experiment. Let's say £10: this is now their money to keep. The subject (player A) is informed that the computer will ask another randomly chosen, anonymous player (player B) if they would like to transfer some or all of their £10 to player A. Why would player B do this? Well, because, as both players know, any amount that player B transfers to player A will be tripled and this will be added when the money hits their account. Player A will then be asked if they want to transfer any money to player B – remember that everything in the game is completely anonymous.

So this game has given rise to some fascinating studies. It is very interesting to see whether people act in both player's best interests, where player B gives their whole £10, resulting in player A having £40, and player A then giving £20 back to player B. Or whether, as economic theory used to predict, that player A should just keep whatever is given to them by player B and give nothing back.

Paul Zak led experiments that utilize the trust game. During one of these experiments player A is left with an average of $14 (in America) and player B with $17. They found that there was a 'dramatic and direct correlation between a person's level of oxytocin and her willingness to respond to a sign of trust by giving back real money' (Zak, 2012). There was even a correlation between the size of the transfer and the size of the recipient's response (more money meant more oxytocin). This oxytocin has a lot of impact, as we have explored in Chapter 11.

In another study, reciprocal altruism was studied using the 'prisoner's dilemma' game. For this, fMRI scanners were used to look at women's brains as they played the game, a version based on the classic scenario formalized by Albert W Tucker, a version of which is:

> Two members of a criminal gang are arrested and imprisoned. Each prisoner is in solitary confinement with no means of speaking to or exchanging messages with the other. The police admit they don't have enough evidence to convict the pair on the principal charge. They plan to sentence both to a year in prison on a lesser charge. Simultaneously, the police offer each prisoner a Faustian bargain. If he testifies against his partner, he will go free while the partner will get three years in prison on the main charge. Oh, yes, there is a catch... If *both* prisoners testify against each other, both will be sentenced to two years in jail.

The results showed that when there was mutual cooperation, areas of the brain involved in reward processing were activated: nucleus accumbens, caudate nucleus, ventromedial frontal cortex, rostral anterior cingulate cortex. The experimenters proposed that the activation of this neural network positively reinforces reciprocal altruism... trust is being built! A later study showed that when cooperation wasn't reciprocated – so a version of when your partner testifies against you – the amygdala and insular show increased activity and people report feeling angry, disappointed and irritated.

One study suggested that decisions about trustworthiness may involve both deliberate and emotional evaluations. The experiment involved participants going into an fMRI scanner and being shown pictures of faces. The first task they were then asked to perform was to indicate whether the face looked trustworthy or not. The second task involved the participants indicating whether the face looked like that of a high school student or a college student. This is known as performing an unrelated cognitive evaluation. Their brains were scanned while they were doing these tasks. There was a little bit more to the study design to make it really meaningful. The results showed that the amygdala was activated more when people saw the untrustworthy faces compared to when they saw the trustworthy ones. The same was noted with the right insula and fusiform gyrus. In another study that looked at people with lesions in the amygdala it was found that those people had impaired judgement of untrustworthiness. So these two studies tie together and help build a reliable picture. The other task that the subjects performed gave a contrast frame. This enabled the researchers to see that the right superior temporal sulcus was activated more when judgements about trustworthiness were being made rather than judgements about age. The orbitofrontal cortex was also involved.

The suggested conclusion from these experiments is that visually relevant information is processed by the superior temporal sulcus. Then the amygdala and orbitofrontal cortex create the emotional response. The amygdala appears to be responsible for the rapid and automatic emotional responses. On the other hand, the orbitofrontal cortex appears to contribute only in the context of a particular conscious evaluation. The emotional response people experience could include both cognitive and somatic changes. We tend to say that 'we have a feeling about someone' as a result.

Our final study in chis chapter looks at the effect of feedback on behaviour. The experiment involved participants making decisions about whether or not to trust trading partners after reading descriptions about their life events that were either slanted towards being praiseworthy, neutral or suspect moral characters. Normally the striatum is involved in us learning and adjusting future behaviours based on reward feedback. In this experiment the participants had to make risky choices and, very interestingly, they persisted in making more risky choices with the partner who had a praiseworthy character. The way the study was designed meant that each of the partners (praiseworthy, neutral and suspect) equally were reinforced. We know from other studies that the caudate nucleus's activation differentiates between positive and negative feedback. In this case it only held true for the neutral partner; it did not at all for the praiseworthy character and only did weakly for the suspect one. The suggested conclusion here is that previous social and moral perceptions can diminish our reliance on feedback mechanisms in the neural circuitry of trial-and-error reward learning.

What can I do with a client this afternoon?

- Raising the awareness of the importance of trust and the implications of trust can help your clients deal with various situations.
- Ask your client how they feel about trust and being trustworthy. Explore how they think people naturally behave.
- Task your client to make a list of the top 10 people who they want to feel they are trusted by. Then suggest they make a list of all the ways they can build trust with those individuals. For example, if one of those people is a direct report of theirs at work perhaps your client will want to choose one thing a week for four weeks that they will tell their colleague that they will do and by when. Then (and here is the revolutionary part for some people) they ensure that they do it.

In a training event with MDs once, an intelligent man questioned whether we could build trust. Here we take the responsible approach that you can and should. While the understanding of what trust is may develop further, currently it is fair to suggest that the neural networks that are laid down over time are important. We are in control of taking actions that others view as trustworthy – and so we are in control of building our reputation as a trustworthy individual.

Fairness

What is it?

Definitions of fairness include an element of being legitimately sought, or achieved, being proper under the rules, and free from bias and dishonesty, along with many more. Most people have a sense of fairness that is socially and contextually programmed in.

Why is it important for coaches?

We do not like it if we don't perceive things to be fair. Our response can be a big one and can have lots of knock-on effects. Being aware of this inbuilt response as a coach is very important. Taking your questions in a certain direction is valuable if you suspect the perception of fairness to be a possible cause of challenge. There may also be times when you will want to explain the neuroscience of the fairness or unfairness response to your client so that they can help others or manage situations more effectively.

What is the underpinning neuroscience?

The anterior cingulate cortex is involved in the process of evaluating whether outcomes are met or are different from expectations. It is also involved when we suffer social losses, such as being rejected or being treated unfairly.

The insula has been shown to light up when people are averse to inequity. People have rejected monetary rewards if they think they are unfair relative to the rewards of others. Research is leading us to believe that, in order for people to accept unfairness, the prefrontal cortex (PFC) is very important. At the moment it looks as if the cognitive PFC quiets down the emotional gut-feeling part of the brain and overrides it.

Productivity takes a hit when people perceive unfairness, because there is decreased activation of the reward centres in the brain and the emotional parts of the brain are disrupted along with the cognitive areas!

What are any interesting studies in this area?

In the area of fairness within social behaviour there is an experimenter's game that is often used. It is called the ultimatum game. The way that humans play the ultimatum game flies in the face of old economic theory. We would expect people to behave in a way that is self-serving. If I offer you £2 (no strings attached) then I would expect you to take it. However, if you found out that someone had given me £10 and I was only choosing to share £2 with you then it becomes a different matter. The ultimatum game has been played all over the world in different currencies, with different amounts, with people of different cultures and the basic results stay the same. Unfair offers are rejected.

A study by Tabibnia and Lieberman (2007) illustrated some important points. Today many organizations offer monetary incentives to try to motivate their employees and then wonder why things don't go exactly to plan. We know from the field of social cognitive neuroscience that human behaviour is far more complex. This study showed the importance of fairness. In the study they saw that fair offers led to both higher happiness ratings and also increased activity in several reward regions of the brain when compared with unfair offers of the same monetary value. Specifically, 50 cents generated a bigger neurological reward than $10 – which sounds crazy until you know that it was 50 cents out of $1, and $10 out of $50.

In another study, this time by Singer *et al* (2006), male and female volunteers played an economic game. Partners played either fairly or unfairly. The volunteers had their brains scanned with fMRI while watching the partners receiving pain. Both men and women exhibited empathy-related activation in the fronto-insular and anterior cingulate cortices towards their partners who had played fairly. In men this response was quite different towards those partners who acted unfairly. The empathy response was reduced and there was an increased activation in the reward-related areas of the brain.

Another study with Tania Singer *et al* got men and women to play an economic game, set up to play either fairly or unfairly. During the second part of the experiment the participants' brain activity was measured using fMRI. For this part of the study the participants were able to see their pair receiving pain. In both men and women the fronto-insular and anterior cingulate cortex was activated, signalling empathy, perhaps, towards the fair players. When pain was inflicted upon the unfair players, however, the empathy-related responses were significantly reduced in men. Instead, there was activation in the reward-related areas of the brain, perhaps signalling a desire for revenge. So the consequences of what is considered fair are significant, especially for men.

In another study participants had their electroencephalogram (EEG) recorded while they played an ultimatum game. They were the recipients during the game and so responded to both fair and unfair offers from other human (so not a computer) proposers. The participants also rated their concern for fairness while they went through this process. Currently it is believed that something called medial frontal negativity (MFN) is linked to the anterior cingulate cortex (ACC), which we know is involved in unfairness. This experiment showed that MFN was more pronounced after unfair offers were made than when fair offers were made. This was even more the case for the individuals who self-reported high concerns for fairness. This research suggests that MFN reflects both whether outcomes match expectations and also whether the process by which these outcomes occur match social norms. The research also suggested that people in bargaining situations are concerned with the fairness of an interaction, not just the outcome for themselves.

Obviously this is only one study, and one that puts people in an artificial situation. So this research should not be extrapolated to mean that every person who has to negotiate as part of their work would have challenges. This is not what the research is saying.

In a further study that expands our understanding of the MFN's response within the social brain, the participants in the study played a classic competitive gambling game where in order for one person to gain, the other had to lose. EEG was employed to measure the neural activity of the individual's perceptions. The MFN showed a significant gender difference. When a female player's opponent lost out in the game the MFN was elicited in them. This can be interpreted that they categorized this occurrence as negative, even though it was to their financial benefit. The same thing did not happen in men.

At a fundamental level, how fairly an individual perceives a situation – be that a work scenario, a family incident or an issue with friends – can impact the outcome of that situation. One study looked at the neural segregation of objective and contextual aspects of fairness. They used a modified version of the ultimatum game and varied the social context. They induced a bias in the participant's acceptance of objectively identical offers and used fMRI to explore correlations between objective and contextual aspects of fairness and the brain's response to these. Their findings show that objective social inequality is tracked in the posterior insula cortex. This objective inequality is integrated with social context in posterior and mid-insula. This is consistent with a fairness construct that is adapted to the social environment. The researchers propose that this might help to explain why we see seemingly inconsistent responses to things that happen. Sometimes people seem to let unfairness slide, and it is not an issue. At other times these same people find a similar (objective) situation to be completely unacceptable. Being aware that fairness is processed by integrating objective and contextual aspects sheds light on why this might be.

An experiment from a different angle looked at how we limit any selfish motives we may have and how we implement fair behaviour. The experiment doesn't address the whole big question (there is a lot to this area in brain terms) but does offer some insights. The study documents the results of disrupting the right dorsolateral prefrontal cortex (dlPFC) (using low-frequency repetitive transcranial magnetic stimulation). The researchers found that this reduced the participant's willingness to reject unfair offers. This indicates that the right dlPFC plays an important role in implementing fair behaviours. Could this mean that if our PFC is not working optimally for whatever reason that tendency to behave fairly may be compromised? Perhaps.

Our final study in this chapter is an uplifting one. It explores the positive emotional impact of fairness. The study examined the neural responses to fair and unfair offers and self-reported happiness levels while controlling for monetary payoff. This is important because, unsurprisingly, the mental processes underlying preference for fairness – and for preferring more money – need to be separated. The clever researchers looked at what happened when participants were given unfair offers that were of equal monetary value to fair offers. They found that high happiness ratings were associated with the fair offers and activation of several reward regions in the brain. They also found that the tendency to accept unfair proposals was accompanied by increased activity in the right ventrolateral prefrontal cortex. This part of the brain is involved with emotion regulation. There was also decreased activity in the anterior insula (involved in negative affect). So, taken all together, we can believe that fairness is hedonically valued. Also, that when we have to tolerate unfair treatment for material gain our brain is going through processes similar to suppressing negative effect.

What can I do with a client this afternoon?

The application of this depends heavily on context. Here are some examples, though:

- If a client has identified a lack of perceived fairness at work then being equipped to have a conversation about the impact that has on people (even citing research) could be a great first step.

- If a client is managing others (in any setting, from business to children to a social group) they will need to be aware of creating transparency and having clear frequent communication.

- Creating clear expectations is another very important thing that you can both use with your client (contracts, agreements, check-ins), and also to encourage your client to evaluate how effectively they are using fairness in their various interactions with people.

Organizations need to be very mindful of our need for perceived fairness. Often the response by individuals will not be a public one if they detect unfairness. As a leader or a manager you almost need to develop a second sense for what could possibly be interpreted as unfair, and pre-empt it either with great communication (if it isn't really unfair) or by making adjustments. Another approach could be to invest in a culture of transparency and trust so that people feel able to voice when they think things are unfair – so that they can be openly addressed.

As a manager you could consider sitting down with your team and drawing attention to all the fair things that already exist within the organization. Sharing how decisions are made for things that are important to them – for example promotions, pay rises, who gets to sit where in the office, etc – can help reduce any unconscious tendencies.

A slightly left field, but very valid, application of the concept of fairness is to consider how an organization is positioning itself to the outside world. An organization whose activities are perceived as unfair, for example the actions Nestlé took many years back in distributing baby milk for free in countries where people could not then afford to buy it, took a long time for people to forgive.

Loneliness

What is it?

Some definitions of loneliness say that it is the state of being alone in solitary isolation, a universal human emotion or a complex set of feelings encompassing reactions to unfulfilled social needs. In this context, loneliness can definitely be experienced even when physically with others. In the terms through which we will explore it here, we can think of it as social connectedness. It is subjective, so a person may have a family, friends and colleagues yet still feel lonely: they can still have a sense of an absence of connectedness. At times it is normal to feel a fleeting sense of loneliness, to feel 'out of it' for example at a party or social gathering. It is not uncommon to feel on the periphery of things from time to time. The challenge, and our focus here, is on when those feelings are prolonged: chronic loneliness. From what we know of the brain, this is likely to have a big impact on individuals.

Why is it important for coaches?

When people are lonely they can struggle in many areas of their life. Their performance at work is likely to suffer (cognitive abilities are affected). They are likely to be less happy at home. Their friendships and relationships could deteriorate. Their health is likely to suffer and self-restraint is affected. Since people can feel lonely even when they seem to have lots of people around them whom they have good relationships with, it is important to be mindful of this.

It has been shown that individuals suffering from chronic loneliness had elevated salivary cortisol levels throughout the course of a day. This suggests that there is increased activation of the HPA axis.

Cacioppo *et al* (2009) suggests that there are three complex factors that are very important when we explore the effects of loneliness. The first is how vulnerable we are to social disconnection. Genetically we are predisposed to need a certain amount of social inclusiveness. Linked to this, as is frequently the case, the environment we grew up in and continue to exist in has a big role to play with how our DNA expresses itself. So there is a degree of where our client is at – that has been predetermined by the time we work with them. Second, there are all the factors that affect a person's ability to self-regulate how they respond to feelings of isolation. An important component here is not just the outward response that people have but the internal one as well. Third, our mental perception of what is happening. How we interpret what is occurring is obviously also fundamental.

What is the underpinning neuroscience?

Dr Ryota Kanai from UCL Institute of Cognitive Neuroscience says that there is a neurobiological basis for loneliness and that through training people may be able to improve their social perception and become less lonely. After doing 108 brain scans Kanai et al (2012) found that the lonely people had less grey matter in the left posterior superior temporal sulcus. This area is implicated in basic social perception. In a very simple test of perception – judging misaligned eyes on faces and whether the people were looking left or right – the lonely people found the test to be much harder.

This is not the entire story when it comes to the topic of loneliness, but another interesting insight into the correlation between what is happening in our lives and the plastic nature of our brain.

What are any interesting studies in this area?

It is worth bearing in mind a study that showed it was possible to *create* feelings of loneliness in people. This study took a group of individuals and manipulated their perceived levels of loneliness with the help of hypnotherapy. The psychiatrist doing the hypnosis used a precise script to reconnect the participants to memories that would make them feel either very connected or very lonely. They found that they could fundamentally alter the responses given to the UCLA Loneliness Scale test. It would be unethical to leave people in a place where they felt lonely. It does open up questions for potential help, though. What would happen to the brains of the individuals made to feel lonely? What would happen to the brains of the individuals made to feel connected?

A study by Lieberman et al (2003) – 'Does rejection hurt? An fMRI study of social exclusion' – really highlighted the effect of social pain on the brain. In this classic experiment individuals were scanned while they played a virtual ball-throwing game with others (or so they thought). After a while they were excluded while the people they played with kept throwing it between themselves. They found that the dorsal anterior cingulate cortex was activated more when they were excluded. The participants also reported feeling distressed. This is a similar response we see in the brain to when someone is in physical pain. The right ventral prefrontal cortex (vPFC) was also active during exclusion. The more it was activated the less distress people reported. This, along with the ACC changes mediating the right vPFC-distress correlation, leads us to think that the cognitive PFC is involved in regulating the distress by disrupting ACC activity.

In the 1990s John Cacioppo did an experiment using the UCLA Loneliness Scale to separate students into three groups (Cacioppo *et al*, 2000). The 'very lonely', the 'not at all lonely' and the ones in the middle. These students then went through a dichotic listening cognitive test. The experiment was a clever one utilizing the brain's natural dominance in one, normally the left, hemisphere. Each student wore a headset and each was trying to identify a sound (like 'wu' or 'ha') that came into one ear, while trying to ignore the irrelevant sound coming into the other ear. Without any additional instructions people showed right ear preference (as expected). When told to focus on using their right ear each group did so with similar success. It got interesting when they were asked to focus on their non-dominant ear. This takes more effort. The group with normal levels of loneliness and those who weren't at all lonely still did well on this test. However, the lonely ones didn't perform as well. They struggled to impose this conscious control and were less accurate in their results.

A couple of related experiments that produced fascinating results were led by Roy Baumeister (Baumeister *et al*, 2005). They found that when people were manipulated to believe that they were likely to end up alone (versus a happy future with significant others) they persevered in drinking more of a concoction. This concoction was positioned to them as being unpleasant in taste but healthy and nutritious. The related study looked at how well people would be able to regulate a behaviour that felt good but wasn't good for them. This time individual participants were told that they would need to make small groups with people who liked and respected each other. After some mingling time they were asked who they wanted to work with. Some individuals were then told that unfortunately no one had selected to work with them, but that they could just do the next part of the task alone. The other individuals were told that because everyone had wanted to work with them, and it was impossible to have such a big group, they would have to do the next part of the task alone. They were told that they were doing a taste test and were presented with 35 bite-sized chocolate-chip cookies in a bowl. They were asked to test them by eating as many as they needed to make an accurate judgement of taste, texture and smell. The disconnected participants (the ones no one wanted to work with) ate an average of nine. The connected participants ate roughly half of that. The first group also reported that the cookies tasted better... but still only mediocre!

The neural correlates for this relationship between social exclusion and self-control were explored with the help of a magnetoencephalography (MEG) investigation. MEG was used with the participants while they did a task where they had to solve 180 medium-difficulty mathematical problems – this was designed to assess their self-control. Some participants were primed to feel socially excluded while the others were in the control group. The socially excluded people showed less activity in the occipital and parietal cortex and the right prefrontal cortex. The results suggest that social exclusion interferes with the executive control of attention. This can be noticed when people are performing specific components of cognitive tasks.

Considering what can be done to help people who feel lonely is a complex task. An experiment that gives us another important insight into what is actually happening was carried out at the University of Chicago. Volunteers underwent fMRI and were shown photographs during this process. Some of the photos were of people, some of objects. Many of the photos were chosen to evoke a positive or negative emotional impact. After the scan, the participants' levels of loneliness were measured. The people who were not lonely showed a greater response to positive pictures of people than positive pictures of objects. There was greater activity in the limbic region, specifically the ventral striatum (known as a reward centre). It is very interesting that the pictures of people had different effects on the lonely versus the non-lonely people. We can draw from this that the answer is not as simple as just putting lonely people in situations with other people (if we hadn't already guessed that!).

Equally interestingly, when the lonely people were shown a negative photograph of people they paid more attention to this than the negative photograph of an object. Their visual cortex and temporoparietal junction showed increased activation.

In the classic experiment where one person acts as 'proposer' and the other as 'decider' and the proposer is given £10, lonely people performed differently to the non-lonely. The proposer gets to choose how to split the money with the decider. The decider either rejects the proposal (in which case neither person gets any money) or it is accepted in which case both get to keep the money. Fairness is key to this game. In this version of the game the propositions were rigged so that half of the offers were unfair (less than £3 proposed). Interestingly, the lonely people accepted more unfair offers than the non-lonely people. They were comparable in their rating of the offers (so they recognized that some of the offers they were accepting were unfair) but accepted them anyway.

Another study looked at the mechanisms by which loneliness might have a negative effect on health. It focused on health behaviours, cardiovascular activation, cortisol levels and sleep. In this particular study, of which there were some limitations (there almost always are), they found that both cardiovascular activation and sleep function were compromised in lonely people.

What can I do with a client this afternoon?

Social connectedness is key to people's performance in life. Therefore, assessing and investing in increasing this can be very valuable. Consider:

- exploring what loneliness really is with your client;
- identifying ways they think they could cushion against it;
- taking the UCLA Loneliness Scale test;
- looking at undesirable behaviours and exploring whether increasing social connectedness would help;
- supporting your client to improve their social perceptiveness;
- drawing up a plan to increase connectedness;
- helping your client to evaluate their current relationships – the quality rather than quantity of them.

CONCLUSION

This book is the first stepping stone of several in any coach's journey to really understanding the neuroscience that is useful to us. It has introduced you to the brain areas that are of most interest and why that is the case. Then we concentrated on the chemicals that will help you to understand the thoughts and actions of your client. Setting aside the neuroanatomy and neurobiology we looked at the foundational concepts that would crop up frequently as we moved forward, including the pivotal neuroplasicity. Drawing on our neuroanatomy we took a different approach to widen our understanding and looked at brain networks. Like watching a good play, we paused for an intermission – where we played with the quantum brain. The final chunks of our exploration were focused around the classic and not-so-classic areas of coaching, where there have been fascinating studies performed that can strengthen and focus our work.

If you have been interested, inspired or excited by anything that neuroscience is offering the world of coaching I would encourage you to take the next step. We have the opportunity to be more credible and more evidence-based and subsequently serve more people better.

By offering insights from neuroscience as an underpinning to coaching we don't have a static model that we are limited by. This is exciting because the field is growing and developing at such a rate it is likely that a revision of this book in five years' time would be double its size! When coaches walk into our Neuroscience for Coaches programme my aim for them is to develop their thinking. There are ways to do that – and encouraging playing with ideas, a safe environment in which to explore ideas and a variety of different sources have proved good starting points. Graduates of the programme are invited to continue to stay up to date by joining the N4C Club. This dedicated group of coaches will grow and explore the applications of neuroscience to our field. My hope is that together we can be pioneers in shaping this for future generations.

FURTHER READING

There are many neuroscientists, psychologists and other interesting people who we choose to keep a close eye on. The research they do, papers they publish and books they write are of high quality and really worth exploring. Check out **www.neuroscienceforcoaches.com** to find out the latest list of these people and resources.

Neuroscience for coaches
For free resources and details of programmes please check out:
www.neuroscienceforcoaches.com

GLOSSARY

action potential The electrical spike that occurs within neurons by way of communicating with other cells, often referred to as 'firing'.

adrenaline Well known for its role in the fight-or-flight response and as being part of our emotional response to things.

amygdala The almond-shaped amygdala is deep within the brain connected to the hippocampus. It is heavily involved in emotional regulation.

anterior cingulate cortex (ACC) The area of the brain primarily responsible for conflict or error detection.

axon A nerve fibre that conducts electrical impulses.

basal ganglia There are billions of basal ganglia, located deep inside the brain and these are key to storing routines, repetitive behaviours and thoughts – so are associated with routine behaviours; habits.

cerebral cortex The outer layer of the brain.

cognition Mental processing, including attention of working memory, problem solving, decision making and reasoning.

cognitive social neuroscience The interdisciplinary field that seeks to understand how biological systems, like humans, implement social processes and behaviours. It uses imaging tools that cognitive neuroscientists favour, such as MRI scanners.

cortisol A hormone that gets released during times of stress. It does several things, including increasing levels of blood sugar and suppressing the immune system.

dendrites The branches at the cell-body end of a neuron that propagate the electrochemical stimulation received from other cells.

dopamine The neurotransmitter involved in how interested we are in things, reward and pleasure networks, regulating emotional responses. This is important in learning.

electroencephalography (EEG) The process where people wear what looks like a swimming hat with wires that records electrical activity from the scalp.

emotions Different definitions in different fields, perhaps useful to consider here is an experience that correlates with a particular physiological activity pattern.

episodic memory The memory of autobiographical events, they have an emotional component to them and may trigger behavioural changes.

excitation The chemical process that makes a neuron more likely to fire an action potential.

feelings The psychological definition is normally the conscious subjective experience of emotion.

fight-or-flight response   Typically when a presumed threat is present the sympathetic nervous system floods the body with adrenaline and noradrenaline, which raise the heart rate, breathing rate and prepare the body to 'fight' or run away.

flow   The term coined by Mihaly Csikszentmihalyi for a specific state of effortless concentration, enjoyment, energized involvement and success in the progressing of the task with which you are involved.

frontal lobes   The area at the front of the brain, rich in dopamine-sensitive neurons. Very important in higher mental functions, including executive functions.

functional magnetic resonance imaging (fMRI)   A type of scanning process used predominantly in research fields that measures brain activity by detecting changes in blood flow. It is similar to MRI (used mostly diagnostically in the clinical world) except that it uses the change in magnetization between oxygen-rich and oxygen-poor blood.

glial cells   Cells that maintain balance, form myelin and provide support and protection for neurons.

Hebb's law   The theory that is summarized as 'cells that fire together, wire together' and is the mechanism for neuroplasticity.

hemispheres   The brain can be seen to have two sections divided by a groove, each section is called a hemisphere. They are connected by the corpus callosum, a large bundle of nerve fibres.

hippocampus   Brain area involved in memory storage, including emotional memories.

hypothalamic-pituitary-adrenal (HPA) axis   The influences and feedback processes between the hypothalamus, pituitary gland and adrenal glands. Forms part of the neuroendocrine system that controls responses to stress.

hypothalamus   An almond-sized structure buried deep inside the brain. It is responsible for regulating several things and controlling their linked behaviour, such as eating, sexual behaviour, sleeping, body temperature, our emotions and the secretion of hormones.

inhibit   In cognitive terms it is the mind's ability to tune out irrelevant distractions.

insular cortex (or insula)   Enables us a degree of awareness and is involved in our ability to be 'in tune' with ourselves.

limbic system   A controversial term (some neuroscientists think it should be abandoned) used to refer to a group of brain structures, including the hippocampus and amygdala, that are involved in a variety of things including emotion, behaviour and long-term memory.

mind   Possibly arising from brain functions; considered the cognitive faculties that enable consciousness and thinking.

myelin   The electrically insulating material that forms a sheath around the axon of a neuron, essential for proper functioning.

neocortex   The outer layer of the cerebral cortex, six layers thin, involved in higher functions.

neuroimaging   The group of technological techniques that enable us to directly or indirectly image the structure, function and pharmacology of the brain.

neuron   Core cells of the nervous system, electrically excitable and transmit information through electrical and chemical signals.

neuroscience   The scientific study of the nervous system. Now an interdisciplinary science that collaborates with other fields.

neurotransmitter   The chemicals that transmit signals from one neuron to another cell across a synapse.

neuroplasticity   The term to describe the adaptable nature of the brain, both on a micro and macro level.

noradrenaline   A compound that has many roles, including as a hormone and a neurotransmitter in the body.

nucleus accumbens (NA)   An area of the brain, part of the ventral striatum that is part of the basal ganglia. It is involved in pleasure, reward, motivation, reinforcement learning, fear, addiction, impulsivity and the placebo effect.

orbitofrontal cortex   The area of the brain located within the prefrontal cortex in the frontal lobes. It is involved in the cognitive processing of decision making and linked to emotion, memory and reward.

oxytocin   Involved in social behaviour, increasing trust, decreasing fear, increasing generosity and also in cognitive functions.

prefrontal cortex (PFC)   Anatomically the frontmost part of the frontal lobes. Involved in executive functions, planning, decision making and social behaviour.

reuptake   The process by which a neurotransmitter is reabsorbed back into a presynaptic neuron. It facilitates the recycling of neurotransmitters and controls how long a signal from a neurotransmitter lasts.

serotonin   Important for mood regulation, appetite, sleep, memory and learning.

striatum   Both coordinates motivation with movement and is also involved with rewards and how we are motivated. It is part of the basal ganglia.

synapse   The structure that enables a neuron to communicate to another cell. The synaptic gap is the area between the neuron and the other cell.

thalamus   You can think of the thalamus as a relay station where nearly all the information filters through to go into the cortex (which is the outermost area of the brain). It is involved in pain sensation, attention and alertness.

triune brain   The now controversial model of the brain consisting of the reptilian, limbic and neocortex.

ventrolateral prefrontal cortex (vlPFC)   The part of the prefrontal cortex that is located opposite the back and away from the midline.

white matter   Composed mostly of glial cells and myelinated axons it modulates the communication between brain regions.

working memory   The limited-capacity memory system that holds information, normally for only a short time. Using it is energy intensive and requires the prefrontal cortex.

REFERENCES

Introduction

Dilts, R (2003) *From Coach to Awakener*, Meta Publications, Capitola, CA
Klein, N (2002) *Time to Think: Listening to ignite the human mind*, Cassell, London

Part One

MacLean, P D (1990) *The Triune Brain in Evolution: Role in paleocerebral functions*, Springer, New York

Chapter 1

Meyer, D E *et al* (2001) Executive control of cognitive processes in task switching, *Journal of Experimental Psychology: Human Perception and Performance*, 27 (4), pp 763–97

Chapter 3

Axmacher, N *et al* (2010) Intracranial EEG correlates of expectancy and memory formation in the human hippocampus and nucleus accumbens, *Neuron*, 65 (4), pp 541–49
Fareri, D S and Delgado, M R (2014) The importance of social rewards and social networks in the human brain, *The Neuroscientist*, 21 February [Online] http://nro.sagepub.com/content/early/2014/02/21/1073858414521869.abstract
Olds, J (1956) Pleasure centers in the brain, *Scientific American*, pp 105–16

Chapter 4

Damasio, A (2008) *Descartes' Error: Emotion, reason and the human brain*, Random House, London
Ogino, Y *et al* (2007) Inner experience of pain: imagination of pain while viewing images showing painful events forms subjective pain representation in humans brain, *Cerebral Cortex*, 17 (5), pp 1139–46

Chapter 6

Choo, W C et al (2005) Dissociation of cortical regions modulated by both working memory load and sleep deprivation and by sleep deprivation alone, *Neuroimage*, **25**, pp 579–87

Chapter 7

Dongsheng, C et al (2013) Hypothalamic programming of systemic ageing involving IKK-β, NF-κB and GnRH, *Nature Journal*, **497**, pp 211–16

Chapter 8

Erickson, K I et al (2010) Exercise training increases size of hippocampus and improves memory, *Proceedings of the National Academy of Sciences of the United States of America*, **108** (7), pp 3017–22

Hebb, D O (2002) *The Organization of Behavior: A neuropsychological theory*, Psychology Press, London

Heijer, T D et al (2011) A study on the bidirectional association between hippocampal volume on magnetic resonance imaging and depression in the elderly, *Biological Psychiatry*, **70**, pp 191–97

Maguire, E A et al (2006) London taxi drivers and bus drivers: a structural MRI and neuropsychological analysis, *Hippocampus*, **16**, pp 1091–101

Chapter 9

Adam, E K et al (2006) Day-to-day dynamics of experience–cortisol associations in a population-based sample of older adults, *Proceedings of the National Academy of Sciences of the United States of America*, **103** (45), pp 17058–63

Dunn, E W et al (2008) Spending money on others promotes happiness, *Science Magazine*, **319** (5870), pp 1687–88

Fjorback, L O et al (2011) Mindfulness-based stress reduction and mindfulness-based cognitive therapy: a systematic review of randomized controlled trials, *Acta Psychiatrica Scandinavica*, **124**, pp 102–19

Chapter 10

Dolan, R J et al (2010) Dopamine, time, and impulsivity in humans, *The Journal of Neuroscience*, **30**, pp 8888–96

Sharot, T et al (2012) How dopamine enhances an optimism bias in humans, *Current Biology*, **22**, pp 1477–81

Chapter 13

Field, T *et al* (2004) Massage therapy effects on depressed pregnant women, *Journal of Psychosomatic Obstetrics and Gynaecology*, **25**, pp 115–22

Chapter 16

Brann, A (2013) *Make Your Brain Work: How to maximize your efficiency, productivity and effectiveness*, Kogan Page, London

Draganski, B *et al* (2006) Temporal and spatial dynamics of brain structure changes during extensive learning, *The Journal of Neurosience*, **26** (23), pp 6314–17

Hebb, D O (2002) *The Organization of Behavior: A neuropsychological theory*, Psychology Press, London

Li, S-C *et al* (2008) Working memory plasticity in old age: practice gain, transfer, and maintenance, *Ulman Psychology and Aging*, **23**, pp 731–42

Chapter 19

Ericsson, K A (1980) Acquisition of a memory skill, *Science*, **208**, pp 1181–82

Hirano, Y *et al* (2008) Effects of chewing in working memory processing, *Neuroscience Letters*, **436** (2), pp 189–92

Miller, G A (1956) The magical number seven, plus or minus two: some limits on our capacity for processing information, *Psychological Review*, **63** (2), pp 81–97

Chapter 20

Langelaan, S *et al* (2006) Do burned-out and work-engaged employees differ in the functioning of the hypothalamic-pituitary-adrenal axis?, *Scandinavian Journal of Work, Environment and Health*, **32**, pp 339–48

Chapter 21

Keysers, Christian (2011) *The Empathic Brain*, Social Brain Press, London

Keysers, Christian *et al* (2003) Both of us disgusted in my insula: the common neural basis of seeing and feeling disgust, *Neuron*, **40**, pp 655–64

New York Times (2006) [accessed 7 January 2014] Cells That Read Minds [Online] http://www.nytimes.com/2006/01/10/science/10mirr.html?pagewanted=all&_r=0

Nummenmaa, L *et al* (2014) Mental action simulation synchronizes action-observation circuits across individuals, *Journal of Neuroscience*, **34**, pp 748–57

Chapter 22

McClure, S M *et al* (2004) Separate neural systems value immediate and delayed monetary rewards, *Science*, 306 (5695), pp 503–07

Chapter 23

Bohr, N (1995) *The Philosophical Writings of Niels Bohr, Vol. 3: Essays 1958–1962 on atomic physics and human knowledge*, Ox Bow Press, Oxford

McTaggart, L (2012) *The Field: The quest for the secret force of the universe*, HarperCollins, London

Ochsner, K N (2002) Rethinking feelings: an FMRI study of the cognitive regulation of emotion, *Journal of Cognitive Neuroscience*, 14, pp 1215–29

Pashler, H E (1999) *The Psychology of Attention*, MIT Press, Cambridge MA

Schwartz, J M, Stapp, H P and Beauregard, M (2005) Quantum physics in neuroscience and psychology: a neurophysical model of mind–brain interaction, *Philosophical Transactions of the Royal Society*, 360, pp 1309–27

Stapp, H P (2011) *Mindful Universe: Quantum mechanics and the participating observer*, Springer, New York

Chapter 24

Baumeister, R F *et al* (1998) Ego depletion: is the active self a limited resource?, *Journal of Personality and Social Psychology*, 74 (5), pp 1252–65

Casey, B J *et al* (2011) Behavioral and neural correlates of delay of gratification 40 years later, *Proceedings of the National Academy of Sciences of the United States of America*, 108 (36), pp 14998–15003

Dweck, C S *et al* (2013) [accessed 7 May 2014] Beliefs about willpower determine the impact of glucose on self-control [Online] http://www.pnas.org

Mischel, W *et al* (1970) Attention in delay of gratification, *Journal of Personality and Social Psychology*, 16 (2), pp 329–37

Chapter 25

Crews, F T *et al* (2009) Impulsivity, frontal lobes and risk for addiction, *Pharmacology, Biochemistry and Behaviour*, 93 (3), pp 237–47

Habib, R *et al* (2010) Neurobehavioral evidence for the 'near-miss' effect in pathological gamblers, *Journal of the Experimental Analysis of Behaviour*, 93 (3), pp 313–28

Hebb, D O (2002) *The Organization of Behavior: A neuropsychological theory*, Psychology Press, London

Smith, K S *et al* (2012) Reversible online control of habitual behavior by optogenetic perturbation of medial prefrontal cortex, *Proceedings of the National Academy of Sciences of the United States of America*, **109** (46), pp 18932–37

Chapter 26

Lowenstein, George (1987) Anticipation and the valuation of delayed consumption, *The Economic Journal*, **97** (387), pp 666–84

Macleod, C *et al* (2002) Induced processing biases have causal effects on anxiety, *Cognition and Emotion*, **16** (3), pp 331–54

Segerstrom, S (2007) Optimism and immunity: do positive thoughts always lead to positive effects? *Brain, Behaviour and Immunity*, **19** (3), pp 195–200

Sharot, T (2012) *The Optimism Bias: Why we're wired to look on the bright side*, Robinson, London

Chapter 27

Berkman, E T, Falk, E B and Lieberman, M D (2012) [accessed 6 May 2014] Effects of Three Core Goal Pursuit Processes on Brain Control Systems: Goal Maintenance, Performance Monitoring, and Response Inhibition [Online] http://plosone.org

Gollwitzer, P and Wieber, F (2010) Overcoming procrastination through planning, in *The Thief of Time: Philosophical essays on procrastination*, ed C Andreou and M D White, Oxford University Press, Oxford

Johnson, M K *et al* (2006) Dissociating medial frontal and posterior cingulate activity during self-reflection, *Social Cognitive and Affective Neuroscience*, **1**, pp 56–64

Lieberman, M D *et al* (2007) Putting feelings into words: affect labeling disrupts amygdala activity in response to affective stimuli, *Psychological Science*, **18** (5), pp 421–28

Spunt, R P and Lieberman, M D (2014) Automaticity, control, and the social brain, in *Dual Process Theories of the Social Mind*, ed J Sherman, B Gawronski and Y Trope, The Guilford Press, New York

Chapter 28

Brown, K W and Ryan, R M (2003) The benefits of being present: mindfulness and its role in psychological well-being, *Journal of Personality and Social Psychology*, **84**, pp 822–48

Dickenson, J *et al* (2013) Neural correlates of focused attention during a brief mindfulness induction, *Social Cognitive and Affective Neuroscience*, **8**, pp 40–47

Jain, S *et al* (2007) A randomized controlled trial of mindfulness meditation versus relaxation training: effects on distress, positive states of mind, rumination, and distraction, *Annals of Behavioural Medicine*, **33**, pp 11–21

Jha, A P *et al* (2010) Examining the protect effects of mindfulness training on working memory capacity and effective experience, *Emotion*, **10**, pp 54–64

Lazar, S *et al* (2005) Meditation experience is associated with increased cortical thickness, *Neuro-Report*, **16**, pp 1893–97

Lutz *et al* (2008) Regulation of the neural circuitry of emotion by compassion meditation: effects of meditative expertise, *PLoS ONE*, **3** (3), pp 1–10

Tang, Y Y and Posner, M I (2009) Attention training and attention state training, *Trends in Cognitive Sciences*, **13**, pp 222–27

Tang, Y Y *et al* (2010) Short-term meditation induces white matter changes in the anterior cingulate, *Proceedings of the National Academy of Sciences of the United States of America*, **107**, pp 15649–52

Chapter 29

Pates, J, Cummings, A and Maynard, I (2002) The effects of hypnosis on flow states and three-point shooting performance in basketball players, *The Sport Psychologist*, **16**, pp 1–15

Chapter 30

Glucksberg, S (1962) The influence of strength of drive on functional fixedness and perceptual recognition, *Journal of Experimental Psychology*, **63**, pp 36–41

Pink, Daniel H (2011) *Drive*, Canongate Books Ltd, Edinburgh

Chapter 31

Cikara, M *et al* (2013) Their pain, our pleasure: stereotype content and schadenfreude, *Annals of the New York Academy of Sciences*, **1299**, pp 52–59

Damasio, A R (1996) The somatic marker hypothesis and the possible functions of the prefrontal cortex, *Philosophical Transactions of the Royal Society of London*, **351**(1346), pp 1413–20

Sanfey, A G (2007) Social decision-making: insights from game theory and neuroscience, *Science*, **318** (5850), pp 598–602

Chapter 32

Scott, D J *et al* (2007) Individual differences in reward responding explain placebo-induced expectations and effects, *Neuron*, **55**, pp 325–36

Volkow, N D *et al* (2004) The addicted human brain viewed in the light of imaging studies: brain circuits and treatment strategies, *Neuropharmacology*, **47**, pp 3–13

Chapter 33

Damasio, A R (2000) Thinking about belief, in *Memory, Brain and Belief*, ed D L Schacter and E Scarry, Harvard University Press, Cambridge MA
Lieberman, M D *et al* (2005) An fMRI investigation of race-related amygdala activity in African-American and Caucasian-American individuals, *Nature Neuroscience*, **8**, pp 720–22
Psychiatric Association Handbook (1994) American Psychiatric Association, Arlington VA
Taylor, Kathleen (2006) *Brainwashing: The science of thought control*, Oxford University Press, Oxford
Winston, J S *et al* (2002) Automatic and intentional brain responses during evaluation of trustworthiness of faces, *Nature Neuroscience*, **5**, pp 277–83

Chapter 34

Bargh, J A *et al* (1996) Automaticity of social behavior: direct effects of trait construct and stereotype activation on action, *Journal of Personality and Social Psychology*, **71**, pp 230–44
Garner, Randy (2005) Post-it note persuasion: a sticky influence, *Journal of Consumer Psychology*, **15** (3) pp 230–37
Hausman, D M and Welch, B (2009) Debate: to nudge or not to nudge, *Journal of Political Philosophy*, **18** (1), pp 124–36

Chapter 35

Daybreak's Stranger Danger, ITV, 4–5 September 2013
Loftus, E F *et al* (2002) Make my memory: how advertising can change our memories of the past, *Psychology & Marketing*, **19** (1), pp 1–23
Loftus, E F *et al* (2013) Misinformation can influence memory for recently experienced, highly stressful events, *International Journal of Law and Psychiatry*, **36**, pp 11–17
Patihis, L *et al* (2013) False memories in highly superior autobiographical memory individuals, *Proceedings of the National Academy of Sciences*, **110**, 20947–52, doi:10.1073/pnas.1314373110

Chapter 36

Covey, Stephen [accessed 6 May 2014] How the Best Leaders Build Trust [Online] http://leadershipnow.com

Zak, Paul (2012) *The Moral Molecule: How trust works*, Penguin Group US, New York

Chapter 37

Singer, T *et al* (2006) Empathic neural responses are modulated by the perceived fairness of others, *Nature*, **439**, pp 466–69

Tabibnia, G and Lieberman, M D (2007) Fairness and cooperation are rewarding: evidence from social cognitive neuroscience, *Annals of the New York Academy of Science*, **18**, pp 90–101

Chapter 38

Baumeister, R F *et al* (2005) Social exclusion impairs self-regulation, *Journal of Personality and Social Psychology*, **88**, pp 589–604

Cacioppo, J T (2009) *Loneliness: Human nature and the need for social connection*, W W Norton & Company, New York

Cacioppo, J T *et al* (2000) Lonely traits and concomitant physiological processes: the MacArthur social neuroscience studies, *International Journal of Psychophysiology*, **35**, pp 143–54

Cacioppo, J T *et al* (2009) In the eye of the beholder: individual differences in loneliness predict neural responses to social stimuli, *Journal of Cognitive Neuroscience*, **21**, pp 83–92

Kanai, R *et al* (2012) Brain structure links loneliness to social perception, *Current Biology*, **22**, pp 1975–79

Lieberman, M D *et al* (2003) Does rejection hurt? an fMRI study of social exclusion, *Science Magazine*, **302**, pp 290–92

INDEX

NB: page numbers in *italic* indicate tables or illustrations

For 'what are some scenarios?' see 'scenarios for helping clients'
For 'what can I do with a client?' see 'coaching roles with clients'
For 'why is it important to me as a coach?' see 'importance to coaches of'

Also available from **Kogan Page**

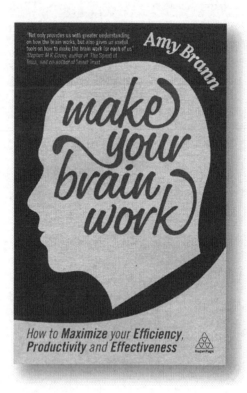